Rocky Shores

BRITISH WILDLIFE COLLECTION 7

Rocky Shores

John Archer-Thomson
and Julian Cremona

BLOOMSBURY WILDLIFE
LONDON · OXFORD · NEW YORK · NEW DELHI · SYDNEY

Dedication

To all the students, worldwide, whom we have had the privilege to teach over the years. Your enquiring minds, enthusiasm and optimism have made our task a joy.

Half-title: Strawberry Anemone.
Frontispiece: A sheltered rocky shore near Dunvegan Castle on the Isle of Skye.

BLOOMSBURY WILDLIFE
Bloomsbury Publishing Plc
50 Bedford Square, London, WC1B 3DP, UK

BLOOMSBURY, BLOOMSBURY WILDLIFE and the Diana logo are trademarks of
Bloomsbury Publishing Plc

First published in Great Britain 2019

Copyright © John Archer-Thomson and Julian Cremona, 2019

John Archer-Thomson and Julian Cremona have asserted their right under the Copyright, Designs and Patents Act, 1988, to be identified as Authors of this work

For legal purposes the Acknowledgements on pp. 359–60
constitute an extension of this copyright page

All rights reserved. No part of this publication may be reproduced or transmitted in any form or by any means, electronic or mechanical, including photocopying, recording, or any information storage or retrieval system, without prior permission in writing from the publishers

Bloomsbury Publishing Plc does not have any control over, or responsibility for, any third-party websites referred to in this book. All internet addresses given in this book were correct at the time of going to press. The authors and publisher regret any inconvenience caused if addresses have changed or sites have ceased to exist, but can accept no responsibility for any such changes

A catalogue record for this book is available from the British Library

ISBN: HB: 978-1-4729-4313-2; ePDF: 978-1-4729-4315-6; ePub: 978-1-4729-4314-9

2 4 6 8 10 9 7 5 3 1

Page layouts by Susan McIntyre
Jacket artwork by Carry Akroyd
Printed and bound in India by Replika Press Pvt. Ltd.

To find out more about our authors and books visit www.bloomsbury.com and sign up for our newsletters

Contents

	Preface	7
1	A fascination for the shore	12
2	Patterns and zones	50
3	Rock pools	90
4	Lichens: a primitive cooperative	118
5	Seaweeds: the banquet that never was	132
6	Stingers, squirts, sponges, mats and worms: the weird and the wonderful	170
7	Molluscs: the mantle of respectability	190
8	Echinoderms: animals with tube feet	222
9	Arthropods: animals with jointed limbs	236
10	Plankton: drifters of the shore	268
11	Attack from air and sea	288
12	Nature's giant compost heap	308
13	Challenges, threats and the future of rocky shores	326
	References and further reading	349
	Abbreviations	353
	Species names	354
	Illustration credits	359
	Index	361

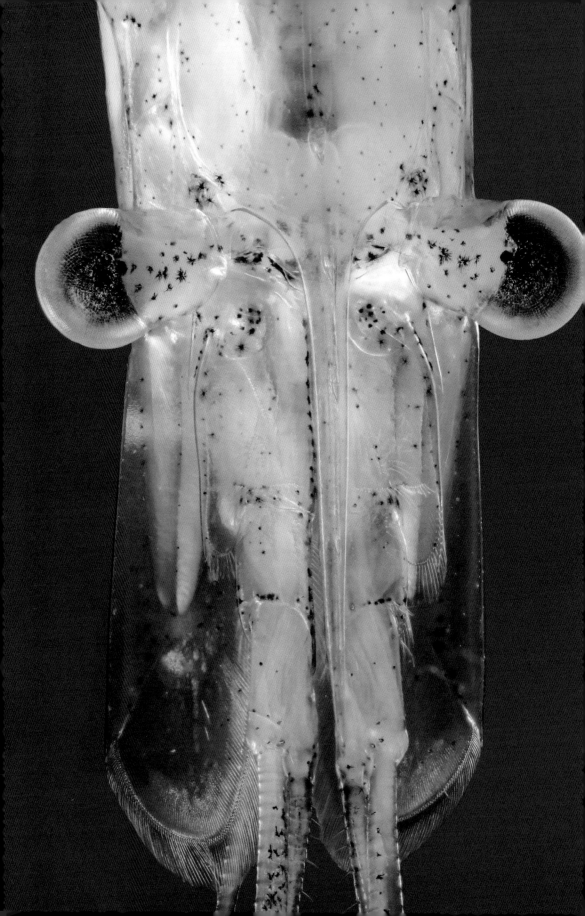

Preface

Where did our love of seashores begin? We both have photographs of us as young children dabbling in rock pools and exploring a variety of other coastal habitats. Our interest in natural history has been a constant throughout our lives, and we are fortunate to have had careers that indulged and enhanced that inclination. Although we were both interested in a broad spectrum of what nature had to offer, it wasn't long before a distinct coastal, and indeed intertidal, bias set in. An abiding interest in the shore and its inhabitants can lead to some unusual lifestyle choices. A fridge full of seaweed and rock-pool specimens is regarded as normal; so-called 'relaxing' visits to the beach often avoid comfort and relaxation on the sand and instead involve a trek to barnacle-covered outcrops in the distance, invariably in inclement weather. Julian's children, now with their own families, comment that wrapping up warm and fighting through the wind and snow to see shore life was what they thought everyone did in the winter.

As serendipity would have it, in the late 1970s and early 1980s, as Julian ran field courses on the rocky shores of Dorset, John was working for the Dorset Wildlife Trust – so we almost certainly worked the same shores, albeit unknown to each other. It was not until 1995 that we met, at Dale Fort Field Centre in Pembrokeshire. John had already worked at the marine centre for 13 years and was deputy head when Julian joined as the new director of studies. It was obvious from the outset that we shared common goals – to educate students of all ages, from a variety of backgrounds, in the importance of conservation and ecology, using the nearby rocky shores, saltmarshes and other coastal habitats as our educational medium. We both valued the rigour of the scientific approach but felt that it was essential at the same time to instil a love and respect for the organisms and habitats we studied.

During the next two decades we worked together, both at the centre and around the British Isles, travelling west to Irish rocky shores, north

OPPOSITE PAGE:
Prawn of the genus *Palaemon*: head viewed from above.

to Orkney and Shetland, and south to Devon and the Bristol Channel. Dale Fort welcomed learners from all over the world – and if they couldn't come to us, we went to them. One long-term and particularly exciting project took us to South-east Asia where we taught 2,000 students and teachers over a 14-year period using their local rocky shores, coral reefs and mangroves as outdoor classrooms; risk assessments had to include heatstroke, mosquitoes, sea snakes and crocodiles.

We retired from our full-time posts at Dale Fort at the same time. John continued with rocky shore and sublittoral survey work, teaching, photography and writing; Julian also with writing, teaching and photography, in particular developing better means of photographing minute seashore creatures.

On one of our trips to the other side of the world, standing on a tropical beach, we first mooted the idea of writing a book together on rocky shores. We had written separately on aspects of shore life before; Julian had produced books such as the *Field Atlas of the Seashore*, while John had taken a rather unorthodox approach and written a guide to the rocky shore through the eyes of a deranged limpet called Larry, namely *The Chronicles of Larry: Volumes 1 & 2*. Rocky shore guides in general often fall into two camps: identification guides that allow interested parties to identify the more common shore creatures but give little information about their lives; and comprehensive, rather academic tomes giving plenty of detail but assuming that the reader is already familiar with the organisms. We hasten to add that this is not intended as a criticism, more a statement of what is. *A Student's Guide to the Seashore* by Fish and Fish (3rd edition, 2011) bucks the trend beautifully by providing help with identification and a wealth of background as well; the authors' names could hardly be more suitable. *The Biology of Rocky Shores* by Little, Williams and Trowbridge (second edition, 2009) is an excellent academic study of the habitat and how it works, with examples taken from all over the world. For identification purposes the *Hamlyn Guide to Seashores and Shallow Seas of Britain and Europe* by Campbell and Nicholls (1994) and the *Collins Complete Guide to British Coastal Wildlife* by Sterry and Cleave (2012) are superb. In 2014 Julian, with John's help, wrote a fully illustrated ecological

Julian on the shore with a Wildlife Watch group member.

Preface

John working with a group of A-level students.

guide to all seashores, but this still had a bias towards individuals with some prior knowledge of ecology.

Our aim in this addition to the British Wildlife Collection is to produce an accessible guide to rocky shores. Chapter 1 looks at what rocky shores are and where one might find them around the British Isles. Hopefully we have provided plenty of historical, geographical and cultural as well as biological background information, as a spur to discovery. The second chapter is the foundation for the rest of the book, an overview explaining patterns and distributions of plants and animals across the shore and the reasons for them. Having set the scene geographically and scientifically, we then devote nine chapters to the creatures themselves, along with their fascinating life histories. Chapter 12 discusses 'nature's giant compost heap', which refers to the strandline of material left by the tide as the water retreats. This part of the shore system might be rather untidy and smelly, and unfortunately these days also bedecked with human refuse, but it is a vital recycling centre for the coastal marine system. Finally, in Chapter 13, we address some of the challenges that the rocky-shore habitat faces, based partly on some long-term data that we have collected over the years with the help of Dale Fort students and staff.

Much of the information in this book is based on our own personal observations of shore life over the decades, but colleagues, teachers, visiting lecturers to Dale Fort and a host of chance encounters along the way have enriched our knowledge immeasurably. We would like to say a profound thank you to all those people for their contributions.

Perhaps more than ever, though, we would like to thank the tens of thousands of students whom we have had the privilege to help, for their enquiring minds, excellent questions, good-natured tolerance of our rather odd sense of humour and ultimately positive view of the world. This book is dedicated to you.

A number of biologists, ecologists and environmentalists we have both worked with over the years also need special thanks: Bill Ballantine (although no longer with us), Juliet Brodie, Peter Browne, Blaise Bullimore, Francis and Anne Bunker, Alec Cooper, Geoff Cooper, John Crothers, Robin Crump, Richard Deverell, David Emerson, John Grahame, Jen Jones, Kate Lock, Phil Newman, James Perrins, Phil Wensley and Denea Wright, to mention but a few. Steve Morrell and Mark and Sue Burton have been good friends and colleagues, and we would like to thank them as well for reading through the final manuscript and making useful comments; it goes without saying that any errors and omissions are our fault alone.

Various biologists and microscopists have inspired us over recent years, particularly Phil Greaves, Carel Sartory and Spike Walker. However, we must particularly thank Mike Crutchley for providing a number of the photomicrographs, typically of planktonic creatures. Klaus Kemp is unique as an artist, creating displays of diatoms only visible under the microscope. We thank him for creating an exhibition-quality slide of 49 diatoms specifically for us to use in the book.

Early one July morning on a peaceful Pembrokeshire shore we met the artist Carry Akroyd with her very buoyant and bouncy dog. For several hours she listened to our enthusiastic ideas for the cover as we pointed out various places and organisms on the shore. As she took copious notes it slowly dawned on us that maybe we had suggested enough material for several book covers. Happily, her illustration is a triumph, as always, and we thank her for creating yet another masterpiece.

We also thank consultant editor Brad Scott for his specialist knowledge, guidance and diligent editing of our text, Hugh Brazier for his copy-editing proficiency and Katy Roper, at Bloomsbury, for her editing skills, commissioning the work and believing in us that it could be done. Finally, huge thanks to Brenda Cremona and Sally Archer-Thomson, for putting up with us, not just during the creation of this book but for all the times that they have provided encouragement and accompanied us on trips to windswept shores. We could not have done it without you.

John Archer-Thomson and Julian Cremona
October 2018

OPPOSITE PAGE:
Low tide on the rocky shore, Ullinish, west Scotland.

A fascination for the shore

chapter one

The summer dawn was about as perfect as it is possible to imagine. We walked with spellbound students over seaweed-covered rocks next to a sea so tranquil the surface was like glass, reflecting the rising sun; only later was this broken by a small pod of dolphins moving through the entrance of Milford Haven. We love the wildness of the shore, the vast array of species to be found there, and, on any day of the year, the uncertainty about what might appear. At any moment, fieldwork could be interrupted by an inquisitive seal popping its head above the water surface, an otter crunching a crab amongst the kelp, or the call of an oystercatcher removing limpets from the rocks.

Seashores are an unusual habitat: terrestrial part of the time, marine at other times, and transitional between these two extremes as the tide ebbs and flows. This applies to both soft-sediment shores and hard rocky ones, but it is on the latter that the most extreme transitions occur. Fine sediments such as mud hold on to moisture throughout the tidal cycle, and even sand remains damp, so environmental change happens quite slowly and the organisms within have time to adjust. By contrast, on rocky shores, surface conditions can change quickly, and creatures have few places to hide unless they can utilise a nearby pool or crevice. As a consequence, many of the creatures found here are exclusive to this thin, rocky ribbon of changeable nature that surrounds much of the British Isles.

Marine systems support an extraordinary diversity of life forms as a result of hundreds of millions of years of evolutionary heritage. Life on Planet Earth started in the sea, probably in an environment similar to the hydrothermal vents found in the deep ocean. One of the

OPPOSITE PAGE:
Rocky shores are a firm favourite with children of all ages.

Rocky Shores

A rocky shore on a very sheltered coast in the Inner Hebrides at low tide, showing an array of seaweeds covering the rocks.

evolutionary pathways from the sea to life on land was via the rocky shore, and that pathway is still open today. More than 30 different animal phyla exist in the sea, three times as many as on the land. This impressive marine richness spills over onto the rocky shore; exploration here can bring observers face to face with at least ten phyla of animals and plants* that are big enough and common enough to enhance any intertidal safari.

Rocky shores defined

In simple terms, a rocky shore is an area of rocky substrate that is covered and uncovered by the tides. First impressions will depend to a great extent on the weather. Warm, windless, sunny days evoke a feeling of calm, when the shore seems a benign environment to explore, but gale-force winds, rain and cold temperatures may render the same shore in a wholly different and altogether more threatening light.

The gradient of the shore can vary considerably, from sheer cliffs to vast expanses of gently sloping rock. Physical processes such as wave action, and the biological activity of resident creatures, model the substrate in many ways. Rarely is the rock surface flat or even.

---

*For convenience we refer to lichens and seaweeds as plants, although – as will be shown in Chapters 4 and 5 – this is not strictly accurate.

Expect to see an infinite variety of shape and size; additionally there may be channels, boulders, pools, crevices and gullies. Life forms differ, depending on the nature of the shore. Some shores may be covered in carpets of seaweed, with splashes of green amongst the brown. Further exploration may reveal small, delicate red seaweeds hiding under the canopy of the larger browns or gracing the sides of crevices and pools.

The rocks may be different colours, sometimes as a result of their geological origin, but invariably the colour palette is embellished with coverings of lichen, especially towards the top of the shore. It may be difficult to recognise animals as such, for their weird and wonderful shapes and appearances may in some instances more closely resemble plants. In fact, the scientific name for the sea anemones and their relatives, Anthozoa, means 'flower animal'. If the seawater brings food and oxygen and takes away waste products and reproductive material there is little point in moving, and the same selection pressures that act to produce plant structures may well come into play for some immobile (sessile) animals. Obvious exceptions to these plant-like animals abound in the form of various snails, crabs, fish and birds, while some other types of animal life might resemble nothing more than coloured crusts or coverings on the rock and seaweed surfaces: sponges, sea-mats and sea-squirts are prime examples. Initial observation of the shore, even from a considerable distance, will confirm that animal and plant coverage varies considerably from top to bottom. The greatest diversity, or species richness, may be found towards the low water mark. Here the structure of the habitat is also more complex, courtesy of the large kelps and other brown seaweeds that dominate. Structural complexity in a habitat can provide many nooks and crannies (or microhabitats) for other species to exploit.

Some shores may lack seaweed cover, except on the lowest part. The middle and upper portions of the shore may be covered with barnacles and limpets, apparently to the exclusion of all else. Closer inspection may reveal tiny little snails hiding in sheltered crevices and even in the remains of dead barnacles. Patches of lichen or specialist seaweeds may invade this animal-dominated terrain if conditions allow.

Why some shores are seaweed-covered and others animal-dominated, with a spectrum of intermediates between these extremes, and why the variety and types of organism differ so significantly with height up and down the shore, are questions fundamental to understanding how rocky shores work – and indeed why they are so fascinating. Essentially, that is what this book is all about.

Rocky Shores

Wave action is a major physical factor limiting life on rocky shores.

In Britain, rocky shores are clearly depicted on the Ordnance Survey Landranger (1:50,000) maps, and reference to these is a good way of assessing the extent of this habitat along a particular stretch of coast; it can extend for many miles or it might be no more than an outcrop of rock on an otherwise sandy beach. The region between the highest and lowest tide points displayed on the maps is referred to as the intertidal or littoral zone. An additional zone is typically found above the intertidal, which is salt-sprayed and hence known as the splash zone. For ease of reference, rocky shores are often divided into upper, middle and lower shore zones, each with its own characteristic species assemblage. Depending on the geology, erosion and location of a shore, rock can be missing in any of these zones. A rocky shore with soft cliffs may have no upper shore at all, as gravel and sand accumulate there instead.

A unique habitat

No two rocky shores are the same. Tides and waves dominate, while the geology influences the form and shape of the shore right down to the presence or absence of microhabitats such as rock pools and crevices. The rock type will determine whether inhabitants can attach to the surface or bore into the substrate. Soft rock will erode differently to hard. As latitude varies, the shores around the British Isles may support the same species or quite different ones, depending on their precise environmental needs. We will see that some species may have 'southern' or 'northern' distributions, often relating to temperature. The actual distribution of a particular species on a shore may also be influenced by temperature gradients up and down it; for example, in western Scotland the Common Sea Urchin *Echinus esculentus* commonly appears on the lower shore, while on the warmer Devon Riviera it remains below the intertidal.

There can be enormous variation in exposure to the intensity of wave action. A wide array of sheltered and exposed rocky shores exists, but even an exposed high-energy shore will have some sheltered areas. The combinations are extensive. Sandy bays have rocky outcrops or headlands, while sheltered shores will invariably have small sediment deposits between rocks; turn over a large rock, and beneath it will be residues of sand or silt. Expect the unexpected. The lower shore can sometimes be entirely sand and consequently devoid of rocky-shore creatures, although large empty shells might provide enough hard substrate for brown seaweeds such as Bootlace Weed *Chorda filum* to attach their long, trailing fronds. Sediments in the lower shore can accommodate marine worm species including the Sand Mason *Lanice conchilega*, and the undersides of large stones might provide shelter for the impressive *Neoamphitrite figulus* worm with its deep red gills and long tentacles. The presence of sand near to rocks can also have a negative effect, producing a scouring action as sand-laden water moves around rocky surfaces, preventing barnacles and mussels from attaching. Fine silt and mud clog the gills of many organisms.

Rocky shores that are exposed to open sea and the action of waves will have a lower diversity than sheltered ones. Seaweeds are less abundant, and the steep rock sections are dominated by animals that can adhere tightly to the surface.

OPPOSITE PAGE:
The approximate locations of rocky shores (shown in red) around the coasts of Britain and Ireland.

Finding rocky shores – a tour of Britain and Ireland

Unexploited rocky shores are one of only a small number of habitats left in the British Isles that might be considered natural and unmanaged. Plants and animals are clearly visible, and most are of limited mobility and a convenient size for handling. Furthermore, the patterns of species' distributions can be seen from a great distance. For anyone wanting to study ecology, this habitat is ideal – and this is one of the reasons why so many students attend seashore courses. Permission is rarely needed to visit the shore as there is a public right of way between high and low water (most of the UK intertidal belongs to the Crown Estate or is held by one of the ancient duchies), although it should be noted that a public right of way to get to the foreshore may not exist.

We were both born within a mile of the sea, but John was the one closest to a rocky coastline at Lyme Regis, while Julian was born in Southampton, a city at the apex of the extensive drowned valley of the Solent. It is a coast of sediment, from fine silts and muds through to sand, gravel and shingle. Unfortunately, except for the occasional large stone and extensive concrete pilings, seaweeds find scant substrate for attachment here. Animals fare little better, although the Common Shore Crab *Carcinus maenas* survives the variable salinities to exist high up the Test and Itchen estuaries, and at low tide the crabs can be seen scuttling across the tidal gravel from the busy bridges full of

Rocky shores and their inhabitants are accessible throughout the year; this photograph was taken in January.

A fascination for the shore

commuter traffic. Fifty years ago, beachcombing around the Solent presented gems such as Sea-mouse *Aphrodita aculeata*, a polychaete worm sometimes found in almost plague proportions but now, with declining numbers, only a distant memory. Weston Shore, which rather surprisingly is on the eastern side of the estuary, has a strandline at the top of the gravel beach consisting of dead seaweed, shells and the detritus of a busy port, but Southampton's nearest rocky shore is across the water on the Isle of Wight.

The northern peak of this diamond of an island projects into the Solent and so is a haven for sediment-loving fauna, but on the southeastern side there are tantalising layers of rocky outcrop exposed at low tide on Bonchurch beach, near Ventnor, with barnacles, limpets and seaweed aplenty. The prize, however, lies at Bembridge, where great ledges create long shallow pools. Once these were an absolute joy for a child on holiday, but in the 1970s this shore was the first in Britain to be invaded by an alien seaweed, Wireweed *Sargassum muticum*, which has the potential to shade out all other seaweeds in the rock pool. From Bembridge the invader has since spread around the British coast, though the saving grace has been that as Wireweed spread west and north the fronds became less dominant and invasive, having far less impact than predicted on the natural communities. In fact, they provide shelter for a host of invertebrates and have increased diversity.

Bembridge Ledges on the Isle of Wight in 1981, colonised by the alien Wireweed.

The south coast

Our tour of rocky shores has stumbled somewhat at Bembridge because, as well as being infamous for the alien landing, it represents, along with the Solent, an important marker on the British coast. If we travel east in our search we will be struggling to find quality rocky shores because of the geology. Along this soft coastline, sedimentary rocks easily erode and the land may slump into the sea. Saltmarshes and sea walls around Portsmouth and Havant protect the muddy shore, making it a haven for wading birds. Eastward again is a succession of shingle ridges and sandy bays, from Hayling Island to Newhaven. The South Downs chalk projects southwards to create the spectacular Seven Sisters and Beachy Head. Although there is a significant amount of shingle, the chalk has been eroded by the sea to form 14km of wave-cut platform. The white cliffs limit access to the shore, with Cow Gap being one place to go rock pooling for anemones, crabs and molluscs, although diversity is reduced by the geology: if the rock is soft then permanent attachment can be difficult, and loose, flat rocks, ideal for hiding under, do not exist. In addition, the upper regions of more typical rocky shores are often missing.

What is excellent along this coast is a wonderful array of marine fossils, including ammonites, echinoderms and corals: creatures that once lived on the rocky shore or whose ancestors do so today. Middle-shore rock pools are enjoyable along this stretch of coast, and these sedimentary rocks are especially good for bivalve boring molluscs, such as the Common Piddock *Pholas dactylus*, which occurs around Beachy Head and uses the teeth on the backs of its two shells to cut into the rock as the animal slowly rotates its body from side to side. It grows larger as it bores into the soft rock, producing a tapered burrow from which it can never escape. This apparently unfortunate arrangement affords the animal protection from predation. Interestingly, Piddocks are bioluminescent, giving off a greenish-blue light, and the Roman commander and naturalist Pliny the Elder wrote that they 'glitter both in the mouth of persons masticating them and in their hands, and even on the floor and on their clothes when drops fall on them, making it clear beyond all doubt that their juice possesses a property that we should marvel at.' The 19th-century scientist Raphael Dubois studied the Piddock along with fireflies in discovering the protein luciferin and how the enzyme luciferase reacts with it to release cold light. Exciting research today shows that when a protein called pholasin (from the

A Common Piddock removed from its rocky burrow.

Rocky Shores

Rocky pools below the white chalk cliffs of Birling Gap, West Sussex.

generic name of the Piddock, *Pholas*) encounters highly reactive chemicals called free radicals produced by white blood cells around human infections the mucus glows green, one of many medical marvels developed from nature.

This part of the coast demonstrates the importance of geology in determining what will survive on a shore. The Piddock does well in this soft rock, but large seaweeds may struggle to maintain a grip on the surface. So we see how the rock type can modify the seashore community. Any folding of the rock will create ridges and depressions that affect the number, size and depth of pools. Depending on their size, loose rocks, when present, may provide shelter for creatures or danger as they move.

Eastern England

Beyond Beachy Head is Dungeness, a monster of a continually growing shingle beach that brings us into St Mary's Bay, with Folkestone and Dover at the eastern end. Here onwards, particularly from Copt Point, an exposed headland replete with mussels and barnacles, is a small oasis of rocky shores before the Isle of Thanet. There are rocks between Margate and Broadstairs with excellent marine fossil forms such as the rare *Uintacrinus* and *Marsupites*. The platforms of soft rock here do not stretch to the cliff and have numerous gaps along the coast. One of these is Dumpton Gap, which should be highlighted because of its Dogwhelk *Nucella lapillus* population. This abundant and widespread snail is found around the coast of the British Isles on rocky shores, although during the decades between 1970 and 1990 there was a noticeable decline in numbers countrywide, especially in north Kent. Surprising then that marine biologist Peter Gibbs from the Marine Biological Association (MBA) found a thriving population at Dumpton in 1992. He identified the individuals as having a genetic mutation, now known as the 'Dumpton syndrome', which stops the development of the penis. This apparent disadvantage in normal Dogwhelk populations proved fortuitous in populations exposed to pollution such as at Dumpton Gap, where females grew non-functional penises, which blocked their oviducts and stopped them laying eggs. Rather than spoil that story here, more will be given later (Chapter 13).

Despite the limited extent of rocky shore present, there is a staggering list of species in *A New Atlas of the Seaweeds of Kent* (Tittley 2016). These, along with the marine molluscs found in Kent, hint at what could be possible if suitable rocky substrate existed here. But now, travelling north across the Thames estuary, we approach mile upon mile of sediment shore. The marshes and saltings of Essex continue around the bulge of East Anglia, with its eroding coast and iconic saltmarshes of Blakeney and Scolt Head. All bereft of any rocky shores. Way beyond the Wash, we can see on the distant horizon the sign for Flamborough Head, in Yorkshire. With few exceptions, this is the first true rocky seashore since we left Bembridge.

Kelp on the chalk shore at low tide, Flamborough Head.

The distribution of species

Throughout this book, when we mention the distribution of a species, some examples are specific, such as 'the southwest coast'. Others may be referred to as 'widespread in Britain and Ireland', although this will only be where there is suitable habitat. We have seen how the Solent is an important dividing line in the east–west split of seashore habitat distribution, but naturalists also talk in terms of a species being 'northern' or 'southern'.

The British Isles lie at an exciting confluence of currents and topography. The North Atlantic Drift brings the Gulf Stream to our west coast, the Bay of Biscay currents are to the south, and the Channel runs up to the North Sea – all located between the Arctic Ocean and the balmy waters of the Mediterranean. Species that are adapted to cold temperatures, which typically live in the north, extend their distribution south so that the southerly limit of their distribution is found in the British Isles.

One of our favourite shores in the world is in Iceland, around the fjord of Blönduós in the north of the country. The rocks are covered in limpets, in particular the Tortoiseshell Limpet *Testudinalia testudinalis*. The shell has a delightful tessellation of colour and pattern, very different to the common British limpets of the genus *Patella*. The Tortoiseshell Limpet is a northern species, and the southern limit of its distribution is reached in Scotland and the north of England.

The beautiful blue Bushy Rainbow Wrack: a warm-water seaweed at the northern limit of its distribution in Kimmeridge Bay, Dorset.

In contrast, when paddling or swimming around rocky shores of the Mediterranean, who could not be delighted by the Peacock's Fan *Padina pavonica*, growing 100mm tall but occurring in great drifts of turf full of waving, triangular fans? This species is a warm-water brown seaweed found in the tropics; in other words, a southern species. In Britain, it just reaches the south coast of Pembrokeshire, although it occurs mainly along the English Channel in suitable habitats.

Naturalists exploring the rocky shores of the British Isles are indeed fortunate to have such an abundance of species around the coast, but a quick trip to the Channel Islands or Brittany will show many other species that have not yet made it across the Channel. This situation is likely to alter as climate change modifies the distributions of our present British fauna and flora.

Kimmeridge Bay: a moving tower and a nodding donkey

The relative protection afforded by the continent and what is left of Doggerland, the ancient land bridge to Europe, plus the sedimentary rocks of the south and east of Britain, results in a paucity of rocky shores between the Isle of Wight and Yorkshire. If instead we move west from the Solent, the geology provides a coast of much harder material. Additionally, the energy generated by the wind on the surface of the sea also has an effect. Our prevailing southwesterly winds cross the vastness of the Atlantic, and as the distance over open water that wind and waves travel (the fetch) increases so does the potential size of the wave. An exposed west-facing shore will receive higher-energy waves than those of the Channel or North Sea by virtue of the differences in fetch.

The geology and the location of the western coasts mean that the diversity of rocky-shore habitats is much greater than in the east. Sediment shores predominate as far as the chalky headland of the Old Harry Rocks near Swanage, but at Durleston Head the designated Jurassic Coast of Dorset begins. The spectacular limestone cliffs here also mark the first great place on the south coast where nesting seabirds such as Guillemot *Uria aalge* and Kittiwake *Rissa tridactyla* can be observed. More importantly, rocky shores stretch to the western horizon.

A view from Gad Cliff, looking east towards Kimmeridge Bay and St Aldhelm's Head on Dorset's Jurassic Coast.

Nearby is Kimmeridge Bay, an important reserve managed by the Dorset Wildlife Trust. The sweep of the bay has amazing cementstone ledges that run out to sea on which you can lie down and look into the clear water filled with seaweed and marine creatures. Peacock's Fan grows in great swathes as though on a Mediterranean shore, while the most magical 'blue' seaweed, Bushy Rainbow Wrack *Cystoseira tamariscifolia*, is iridescent in the sunlight, and every rock seems to have a Black Squat Lobster *Galathea squamifera* beneath.

Beyond the sheltered shore of Kimmeridge to the west is a smaller shore, Charnel, which is more exposed to wave action. Walk further and a large expanse of rock, aptly called Broad Bench, projects to the south. From this exposed shore a spectacular view extends east and west for miles.

The popularity of this stretch of coast began in the 18th century with visitors staying in the developing resort of Swanage. In 1830 the landowner, Sir John Clavell, built a folly, Clavell Tower, on top of the cliff to the east side of the bay where people could admire the view. The Victorian author Thomas Hardy was a regular visitor to the bay and is believed to have wooed his first love, Eliza Nicholls, in the tower. Railways afforded people easy access to the coast, and Swanage opened a branch line in 1885 to receive passengers from London. By the end of the 20th century, however, the tower had been closed for years, the desolate ruin in danger of collapsing over the eroding cliff. In 2006 the Landmark Trust took on the challenge, removed the tower brick by brick and rebuilt it 25m back from the cliff edge.

The accessible rocky shore of Kimmeridge Bay is iconic because of both its history and its location. Divers, schoolchildren and students come and look at marine life, while the gorgeous views offered at sunset have it in many landscape photographers' portfolios. As a rocky shore, it does share one problem with the shores east of here: the lack of a typical lichen-covered splash zone. This is due to unstable bituminous shale. Positioned just above the tide line is the richest oil-bearing mineral in Britain, yielding over 80 gallons of oil per tonne of shale. Locally the blackstone, as it became known, was burnt as coal, but the high sulphur levels meant that it had a strong and unpleasant smell. Sir John Clavell exploited the oil for making alum and gas. In 1884 the streets of Wareham were lit with gas produced from the shale, and there were ambitious plans to light the whole of Paris. From time to time – most recently in summer 2000 – the rocky shore used to be seen on fire, and even the periwinkles moved down the shore to

At any age, looking into a new and alien world is always exciting. Here at South Milton, Devon, the upper shore is invariably of sand. Note, in the foreground, the Wireweed that has colonised the pools.

avoid the smoke. In fact the cliff can remain smouldering for many months. Kimmeridge was the first onshore drilling area in Britain, and oil extraction continues today with an oil pump, a nodding donkey, bringing up oil from below the seabed into holding tanks.

The rugged west

Spectacular landforms make the famous coastal sites of Lulworth, Durdle Door and Portland Bill honeypots for geographers and photographers alike. These occur on the hard Portland limestone, west of Kimmeridge's softer oil-bearing shale. Marine fossils abound on this coast. The sweep of Lyme Bay is dominated by the shingle of Chesil Beach, and while there are occasional excellent rocky shores, it is towards Plymouth in Devon that we travel next. The coastline around Torquay and Paignton is now a sprawl of hotels and tourist attractions but their origins lie, like Swanage, in an era of fascination for the seashore. In a cemetery near the sea in Torquay is the grave of William Henry Harvey, often said to be the 'father' of algae, buried here in 1866. Harvey originated from Limerick and travelled the world studying and writing about seaweeds.

There are many superb stretches of coast in this area, but the broad, level shore at Wembury, near Plymouth, has long been a beacon for rock poolers. The land is owned by the National Trust, and a marine centre is run by the Devon Wildlife Trust. The site is located in a Special Area of Conservation (SAC) and a Voluntary Marine Conservation Area (VMCA). Wembury is renowned for the biodiversity and quality of its rock pools, which have featured, more than a few times, on national television. Dozens of rock-pool safaris a year are open to the general public, and it was on one of those in August 2016 that St Piran's Crab *Clibanarius erythropus* made an appearance. This tiny hermit crab was thought to have disappeared from the southwest, possibly because of pollution after the oil tanker *Torrey Canyon* was wrecked between Cornwall and the Isles of Scilly in 1967. It just shows how important these public events can be for monitoring our wildlife.

Plymouth city itself is home to the MBA's Citadel Hill Laboratory, opened in 1888. The society has a world-leading reputation for marine biological research and has produced a number of Nobel Prize winners. The foundation of the organisation lies in a conflict of opinions over the fishing industry. In 1883 Professor Thomas Huxley believed the fish stocks of Atlantic Cod *Gadus morhua* and Atlantic Herring *Clupea harengus* to be inexhaustible, while Professor Ray Lankester, whom we will meet again in Chapter 6, thought otherwise. His reasoning was

National Trust rocky shore at Wembury, south Devon.

that removal of large numbers of individuals from a fish population would upset the balance of nature. In an impassioned speech, Lankester said the only way to prove this was by way of research, and that a society prepared to take this on would be necessary. Luckily a group of scientists answered the call to arms, and the Marine Laboratory of the MBA was established. It is right by the sea, and generations of students of every age have explored the sheltered and accessible rocky shore on its doorstep.

Plymouth Marine Laboratory, Devon.

West of Plymouth lies the rugged coast of Cornwall, home to some of the finest rocky shores in southern Britain – and to tales of illicit goods and the wreckers who used these remote rocky bays and drowned river valleys. The small town of Polperro is one such place, famous for smuggling in the 18th century as well as for the generations of children and families who have swum in the Chapel Tide Pool. Bathing in large rock pools has been a popular pastime in the southwest. It may have started with the health fad for seaweed baths in the early Victorian period, but its popularity today is more likely due to the fact that at low tide the large pools warm up on a sunny day to be slightly above the temperature of open water. With the addition of concrete dams on the rocky shore, the number of artificial pools has been extended on tourist beaches in both Cornwall and north Devon.

Kynance Cove on the Lizard Peninsula is renowned for the serpentine rock, coloured green and red with veins of yellow, while to the north Bedruthan Steps (see overleaf) is a famous stretch of Cornish coastline comprising sandstone and shales. Sections of cliff have broken away to form stacks and rocks amongst sandy sections of beach. The origin of the name is unclear, and the legend that Bedruthan was a giant who used the stacks as steps to cross the bay is thought to be a relatively recent story. Centuries ago, steps were cut in the hard sandstone to descend the cliffs to the shore. This must have been a painstaking task, and illustrates the determination that people had to explore these tantalising environments. Ancient steps are a feature of a number of the westerly cliffs, and they may well have had links to smuggling activities. Smugglers needed sandy beaches for landing their boats, so they were not as active along the north Cornwall coast as they were in the south, where sandy coves are more prevalent.

A mix of rocky shore and sandy bays pervades the north Cornwall coast, as here at Bedruthan Steps.

The wild rocky coast continues northeast towards the Bristol Channel and the record books. Avonmouth in Somerset has the third-highest tidal range in the world after the Bay of Fundy and Ungava Bay on Canada's east coast. North from Burnham-on-Sea and Weston-super-Mare, the estuary narrows and the sediment builds, as does the height of the shore, so that near Avonmouth and the Severn Bridge a maximum tidal range of 15m can be reached, although an average spring tide is closer to 12m. The British Isles exhibit an incredible variety of tidal ranges: Pembrokeshire has a range of approximately 8m; north Cornwall an average of 6m; Portsmouth and the Solent less than 4m; and this decreases to only 2m at Portland. For the smallest tidal range nothing beats the shores around Islay and Jura, where the average is a little over half a metre.

Welsh shores, and Britain's only coastal national park

The Glamorgan Heritage Coast is a stunning 23km stretch of carboniferous limestone between Cardiff and Swansea, and Nash Point has long been a popular shore for locals and tourists alike, with the geology creating a shoreline of giant, tide-swept steps. Beyond Swansea is the Gower Peninsula, a huge wave-cut limestone platform, and the coast here was the first to be designated an Area of Outstanding Natural Beauty (AONB), in 1958. Much of the shore from Swansea through to the projection that is Worm's Head has long been used by marine biologists from Swansea University, where two leading scientists, Professors John Ryland and Peter Hayward, have probably written more on shore organisms than anyone else, both separately and together.

The year 1856 was significant for Pembrokeshire's coast. For one, the Victorian naturalist Philip Henry Gosse went there on his summer holiday and shortly afterwards wrote the book *Tenby: a Sea-Side Holiday*. This may sound trite compared to another of his books, *A Year at the Shore* (1865), but all his works were well written and inspired generations. In fact, a more appropriate title for his Tenby book might have been *Latest Research on Marine Plankton*, much of which helped to complement Darwin's work. But Gosse knew what he was doing to attract and enthuse his audience. The book is immensely readable 150 years later, and his illustrations are exquisitely beautiful and detailed. In *Tenby*, Gosse describes the seven-hour journey on the Great Western Railway across Britain from London, stressing the importance to naturalists like him of this amazing transport. What struck him most on arrival at Tenby were people eating Welsh cakes: '[they] kept munching enormous flat cakes, like pancakes, but resembled short piecrust. This sort of pastry, which looks by no means despicable, is evidently a staple commodity here.' Gosse was an inspirational

A low-tide rock pool at Worm's Head, Gower Peninsula.

Rocky Shores

One of many drawings made by the Victorian naturalist Philip Henry Gosse to illustrate his books, in this case after visiting Tenby in 1856.

Victorian naturalist, and it was after his rocky-shore explorations around Tenby that he developed a recipe for artificial seawater, which enabled the Victorian trend for home marine aquaria (see Chapter 3).

As well as being the year of P. H. Gosse's holiday, 1856 was also significant for being the year the government built Dale Fort at the entrance to Milford Haven Waterway. Projecting into the Irish Sea, Pembrokeshire is surrounded on three sides by water, deeply incised with estuaries, and it is host to the Pembrokeshire Coast National Park, Britain's only coastal example. A path around the county hugs the shoreline of almost 300km of cliffs, islands and a huge diversity of rocky shores. In 1946 the Field Studies Council (FSC) was searching for a centre in which to develop marine education. The fledgling organisation had just acquired three National Trust properties to run education programmes, the first being Flatford Mill in Suffolk. The problem was that there were no Trust properties near the coast. So the empty Victorian fort at the entrance to Milford Haven was bought by the local West Wales Field Society on the understanding that the FSC would take over its running.

Dale Fort was close to being a ruin after the Navy used it during the Second World War as a station for degaussing: measuring the magnetic fields around ships coming in from the Atlantic convoys to help reduce their chances of detonating mines. Dale Fort was also a mine watching station, where WRENs would spend all night looking for mines being parachuted from German aircraft. The job of the

Dale Fort Field Centre, located at the entrance to Milford Haven.

first warden in 1947, John Barrett, was to convert a building that in his own words was 'in a deplorable state of rot and damage' into a marine station suitable for teaching marine biology. At a time of rationing, no tools or paint in shops and a licence required to buy wood, he scoured beaches for driftwood to help build marine labs. By chance a boat was wrecked on a local rocky shore, and from it he extracted nails and wire, and wood to construct furniture. Electricity arrived in 1954, but drinking water was rain washing off the roof (complete with gull droppings).

Barrett was not a biologist, but he taught himself ahead of the students and in 1958 wrote the *Collins Pocket Guide to the Seashore*, the first of a series of books he would go on to write on seashores and the marine environment. For 70 years, Dale Fort Field Centre – the world's first dedicated marine centre – has been at the forefront of seashore education in Britain and has helped in the development of other seashore centres around the world, including in New Zealand and Galapagos. Considerable amounts of baseline data have been collated and published for the surrounding rocky shores, in particular by Dr John Crothers, who worked at Dale with John Barrett during the 1960s.

John Barrett, the first warden of Dale Fort, teaching an A-level student in the library, c. 1956.

Rocky Shores

A view over Skomer Island, a Special Area of Conservation, towards Milford Haven.

Pembrokeshire is also noteworthy for encompassing the UK's first marine-designated Special Area of Conservation (SAC). Stretching from Milford Haven estuary, the SAC extends beyond the horizon to the islands of the Smalls and north to cover much of the Cardigan coast, taking in exquisite rocky shores and offshore habitat for cetaceans. For such a limited area, these shores are some of the most diverse in Britain, from extremely sheltered to extremely wave-exposed. They incorporate a huge biodiversity and as such represent some of the most studied seaweed-covered rocks in history. The marine SAC embraces internationally famous seabird islands including the larger ones of Skomer, Skokholm and Ramsey. Some, like Grassholm, a reserve with almost 40,000 pairs of Gannets *Morus bassanus*, are owned and managed by the Royal Society for the Protection of Birds (RSPB).

Further north, the rocky shores around Aberystwyth have for generations been used by biology students, both from the local university and from all over the UK, for their marine biology field courses. Bangor University also has a long-standing reputation for oceanography and marine biology, with students studying the rocky shores and tidal streams of the Menai Straits.

Marine research

Both the Victorians and the Edwardians created a legacy of demand for marine research and education, and scientists realised there was a need for laboratories near good rocky shores. The year 1892 saw the opening of the Marine Biology Station at Port Erin on the Isle of Man, and the huge demand for visits to this stretch of rocky shore from naturalists, tourists and students led to a relocation to larger premises in 1902. The station was eventually taken over and run by the University of Liverpool in 1919. By 2005, the last students were using the site, and sadly it has since become derelict – but there is a plan to turn it into a marine interpretation centre, a trend we may well see repeated around the coast (see Robin Hood's Bay, p. 48). University cost-cutting has begun to undermine much of the teaching of undergraduates in these emblematic field sites. Students still attend field courses at places like Dale Fort, of course, but it is cheaper now to send students for a week in the Mediterranean than to British shores.

Located on the southern tip of Great Cumbrae, a small island within the Firth of Clyde, is Millport, another iconic marine biology location that happily has a better story to tell. Through much of the 19th century, visitors had been attracted to the Firth of Clyde for its outstanding sheltered rocky shores and wide-ranging variety of seashore life. The eminent marine biologist Sir John Murray returned from the four-year *Challenger* oceanographic expedition in 1876 fired up to develop marine research in Scotland. He bought a 25m boat, *The Ark*, which he eventually moored on Cumbrae in 1884 as a floating laboratory. This formed the Scottish Marine Station for 12 years and was visited and used by many scientists and naturalists. David Robertson took over the running in 1894 (later the museum and laboratory were named after him) and two years later built the Millport Marine Biological Station. By 1914 it had become the base for the newly formed Scottish Marine Biological Association, and it continued to thrive until expansion required a base on the mainland. Moving to the current base at Oban in 1970, the association has developed into the Scottish Association for Marine Science (SAMS), with research carried out worldwide. Millport was then funded by tertiary education from London University and called the University Marine Biological Station Millport. By 2013 funding had ceased, and over a century of marine biological research came to an end. But in 2014 the Field Studies Council breathed new life into the project and continued the running under an environmental charity banner. Research activity at Millport may have declined, but now children and adults of all ages can enjoy the fascination for rocky shores.

Millport Marine Lab, Great Cumbrae, 2016.

The west coast of Scotland has vast expanses of rocky-shore habitat, with the greatest variety anywhere in the UK.

Highlands and islands

Between Anglesey and the south of Scotland good rocky shores are few. The coast of Cumbria is peppered with them, but much of the coastline of northwest England is made up of spectacular sediments, as seen in Formby Dunes or the expanses of Morecambe Bay and the Solway Firth.

As we head out along the Galloway coast, however, the landscape changes. The entire western coastline of Scotland, from the Mull of Galloway to the northern tip of Shetland, has the most extensive rocky shores in Britain. Biodiversity can be exceptional, particularly in some of the sheltered parts of the Outer Hebrides such as the eastern coast of Benbecula. The shores here may superficially resemble those of Pembrokeshire, but they are colonised by more northerly species, requiring cooler conditions. Jagged peninsulas like Kintyre and Ardnamurchan project a blanket of rocky-shore seaweed into the Atlantic; countless islands of every size from a simple seaweed-strewn rock to the magnificence of Harris and Skye are adorned with breathtaking shores. We have been privileged to visit many Hebridean islands, but even those trips have covered a remarkably small sample of coast relative to the vast array of island coastline in total. No other region in Britain provides such a range of shores, from the extreme wave-lashed, barnacle-dominated rocks of the Atlantic-facing shores to the extreme shelter afforded by the many sea lochs lining the edge of the Minch. Some of these lochs have shores so far from the sea that tidal movements are expressed as just a gentle lowering and rising of the water. The rocks here are so dense with seaweed that animal life is almost impossible to see. With no wave action, the large brown seaweeds can grow to almost twice the size as any seen elsewhere; curves develop in the fronds where normally they are straight, and the kelp Oarweed *Laminaria digitata* forms huge unbroken blades instead of the usual digitated ribbons. Otters *Lutra lutra* dip in and out of the water and Grey Seals *Halichoerus grypus* bask on the rocks in impressive numbers. This is rocky shore heaven.

The topography of the land plays a large part in marine biodiversity. Visitors to Scotland's west coast should first spend some time poring over Ordnance Survey maps. The nature of the coast – with so many headlands and sea lochs – creates an unusual marine habitat, a tidal rapid, that is found in few other regions. Imagine a large sea loch open to the sea by a very narrow exit point. Tidal flow through this constriction, both in and out, brings a constant renewal of nutrients and oxygen, stopping for just a short period at low and high tide when the flow reverses. The relentless movement of seawater requires organisms to hold on tight to their substrate, but the rewards for doing so are boundless. One of the most accessible tidal rapids can be seen

Tidal rapids at the entrance to Pool Roag on the Isle of Skye, on a low spring tide. The kelp in the water is flowing from right to left, reversing at high tide.

from the A82 main road south of Fort William at Ballachulish, where Loch Leven empties into Loch Linnhe. An especially unusual example is Loch Obe on Barra, where the narrow entrance is a steep-sided channel almost a kilometre long. Leiravay Rapids and Sponish Rapids near Lochmaddy, North Uist, and Pool Roag on Skye are other superb places to visit.

But tidal rapids are not just associated with sea lochs. With so many islands, inevitably there are narrow straits and channels between them, for example between Benbecula and South Uist. Tides are different on either side of the islands, in this case between the Atlantic and the Minch. The difference creates a tremendous flow of water through these gaps. Filling the shallow depths are unusual maerl beds, almost like coral, composed of calcareous red seaweeds growing in three dimensions within the flowing water. These tidal sweeps can form dangerous whirlpools off island headlands, most famously in the Gulf of Corryvreckan off the Isle of Jura, often described by divers as the most dangerous dive in Britain. In 1947, George Orwell spent some time on Jura seeking inspiration for his novel *1984*. On one of his daily boat trips, his small party became trapped in the whirlpool, and only with the slackening tide did they finally escape the manic water. They were then tipped out of the boat and became trapped on a rocky outcrop, and were eventually

The Straits of Islay, looking towards Jura. The water here has the smallest tidal range in Britain, with an average of just 0.6m. The wreck is the trawler *Wyre Majestic*, which ran aground in 1974.

rescued by a lobster fisherman. Orwell's manuscript was finished three months later.

Close to the western edge of Jura lies the island of Islay, a two-hour ferry crossing from the mainland. There are several important reasons to visit, not least of which is the large number of malt whisky distilleries – all well worth a call, especially as most are located in small bays. The Lagavulin distillery is on a delightful, sheltered rocky shore, and a little further west is a small rocky bay with the Laphroaig distillery. One section of shoreline that is especially attractive, with both malt whisky and kelp forest, is that beside the Caol Ila distillery. Gaelic for 'the Straits of Islay', this modern distillery gives an unparalleled view across the narrow straits to Jura.

During 1977 we were working on various shores in the Outer Hebrides, driving back and forth with students along the lonely, single-track road that connected South and North Uist. The unforgiving terrain naturally slowed the traffic, but on many occasions we were brought to a standstill by lorries overladen with Egg Wrack *Ascophyllum nodosum*, the pile on top slowly slipping off the back onto the road. The alginate industry in the Outer Hebrides was thriving in the 1970s, but when we returned in 1981 with more students the slow lorries and their trailing seaweed were already becoming a rare sight, owing to cheaper imports.

Drying seaweed at an alginate factory. South Uist, Outer Hebrides, 1981.

Rocky shores and post-Ice Age Britain

The Neolithic standing stones of Callinish, Isle of Lewis.

With lower levels of disturbance than many other shores around Britain, those of the Scottish islands have offered up some unique archaeological information about our ancestors.

We may think that our fascination with shores is comparatively recent, but after the last Ice Age, as the land bridge disappeared under water and Britain became an island, the coast provided food for the hunter-gatherers of the time. The population in 8000 BC was less than 20,000, and most inhabitants were on the south and east coasts of what is now England. But we also know that people were living on the rocky shores of Loch Sresort on the eastern side of Rum. This was such an important site that they stayed for 4,000 years, knapping flint and collecting agate and quartz. Most valuable was the unusual rock chalcedony, washed by the tide on the west of the island. This beautiful green and red rock would be brought back to Sresort, which was sheltered, so that it could be worked into hammer stones and sharp blades. Communities used the islands of Islay, Jura, Mull and Tiree, amongst many others, for summer hunting, but no habitation was so permanent and long-lived as that on Rum. Hunters returned from these forays around the coast with fish, seabirds (especially Puffins *Fratercula arctica*) and seals as their main protein sources.

Sea levels were rising at this time as the ice continued to melt. To Scotland's east, the sea level had risen by more than 4m. The human communities that dotted the coast had to move to higher ground as their sites were flooded, and our knowledge of them therefore becomes more limited. In the northern peninsula of Scotland, the land turned into the islands of Orkney and Shetland and was colonised. Between about 7000 and 6000 BC, a tsunami from North America hit these northern islands and a 25m Atlantic wave swept everything from Orkney and Shetland, wiping away fish stocks and shellfish. It completely devastated the north coast, but the impact was felt across much of Britain. One of the key areas of evidence for this and the aftermath is at Maggie Kettle's Loch in Shetland. The significance of this event was not just in the wiping out of much of the population but also in the way it changed how the surviving populations treated the rocky shore. For several millennia it had been life-giving, but now it was to be treated with fear and distrust. As a result, a noticeable change in the location of at least semi-permanent communities occurred: they moved inland.

Since W. G. Hoskins' book *The Making of the English Landscape* was published in 1955, many authors have tried to analyse how our ancestors used the land, but the geographer Nicholas Crane, in his 2016 book *The Making of the British Landscape*, lays down important markers as to how the coast has been vital to our development. Despite the Atlantic tsunami, the rocky shore continued to be visited for food, as it was easy to collect, but more typically from 6000 BC shores were used as 'fast-food restaurants', that is, not as main sources of food. Analysis of sites around Orkney, Northumberland, Portland and Bideford shows middens (piles of waste) full of seabird bones, scallops, oysters, lobsters, mussels, winkles, whelks, limpets and occasional seal bones. In middens on Oronsay in the Inner Hebrides, dated around 5300–4300 BC, were fish such as Angelshark *Squatina squatina*, Thornback Ray *Raja clavata* and Tope *Galeorhinus galeus*, and birds such as terns, Shag *Phalacrocorax aristotelis*, Cormorant *P. carbo* and the now extinct Great Auk *Pinguinus impennis*. These waste heaps were so large that they would have been used as beacons or guides for boats at sea.

Scotland experienced ice over 1.5km thick during the last Ice Age, similar to the depth of ice found in Greenland today. Then, during the postglacial time the land groaned and heaved a sigh of relief as the weight was removed. The west side in particular rebounded and rose. The rocky shores of 8000 BC are now high and dry above the present-day tidal range. These raised beaches are easily spotted all over the west coast, like a second rocky shore above the current one.

The north coast of Skye displays a number of raised beaches, as here around Staffin.

The island of Ireland

Is there anywhere else in the world that possesses a coastline so varied, with such spectacular geology, topography and biodiversity, in as compact an area as the British Isles? The west coasts of England, Wales and Scotland are astonishing, and this high standard is maintained on crossing the Irish Sea. The Giant's Causeway in Northern Ireland, a UNESCO World Heritage Site renowned for its polygonal columns of layered basalt, is particularly revered, but in fact the entire coast of Ireland is remarkable.

Lough Hyne is a sea lough in west Cork, about 5km southwest of Skibbereen. It was designated as Europe's first Marine Nature Reserve in 1981 and, like some of the sea lochs of Scotland, it has a splendid set of tidal rapids where the daily tides flush through the long, narrow entrance where the water is less than 5m deep. In 1886 the Reverend William Spotswood Green made the first record here of the Purple Sea Urchin *Paracentrotus lividus*, a Mediterranean species found more typically on the west coast of Ireland. The urchin has since become one of the most studied organisms in the lough and is currently the subject of a long-running population survey by Dr Anne Crook at the University of Reading. Regular biological studies started in 1923, when Professor Louis Renouf of University College Cork began visiting the lough. The first laboratory, built in 1928, was constructed from an ex-army hut and assembled next to the rapids. During the 1930s, Professor Jack Kitching made annual trips to the rapids, and between 1946 and 1988 he extensively researched the ecology of the lough with his students. In 1996 he wrote the first edition of the concise academic work *The Biology of Rocky Shores* with Colin Little. University College Cork maintains research work here with three laboratories, two of which are named after Renouf and Kitching. If you find yourself in the area, the displays at nearby Skibbereen Heritage Centre are certainly worth a visit.

To the west of Lough Hyne, the tourist routes follow the deeply incised rocky coastline around County Kerry. Long rias or drowned valleys have expansive sheltered shores on their inland fringes, while the exposed cliffs facing into the Atlantic produce magnificent headlands, many over 300m high.

Slea Head on the Dingle Peninsula is one of the most westerly points in Europe. Just a few kilometres east of here is the rather insignificant-looking Parkmore Point, a legendary site to the seashore ecologist since

Lough Hyne, in the area of the Kitching Lab and the tidal rapids.

it features heavily in the landmark publication of 1964, *The Ecology of Rocky Shores*. For every subject or genre there is always a bible, a go-to book, and Dr Jack Lewis's academic account of the dominant plant and animal communities on the rocky shore, filled with diagrams and photographs, is essential reading. Rather than concentrating on the organisms, he takes a holistic approach to zonation and the factors affecting the shores of the British Isles. Our copies are not so much thumbed as loose-leaf files.

Heading north, the dramatic rocky coast of Kerry drops down to the vast estuary of Ireland's longest river, the Shannon. Take the ferry across the Shannon from Tarbert and you enter the very different world of County Clare and the Burren. The wave-cut platforms making up the exposed rocky shores that face the Atlantic are astounding for their growths of kelp and Thong Weed *Himanthalia elongata* on the lower shore. For centuries, local people have collected the seaweed for soil fertiliser along this west coast. Although less so now, kelp is stacked up to 2m high in runs of up to 10m long near the road above the tide zone at places like Kilkee and Quilty. North of Lahinch are the most spectacular and accessible cliffs in Europe, the Cliffs of Moher, rising sheer from the sea to 214m, dwarfing the tiny auks swimming in the sea below. The carboniferous limestone becomes exposed from Doolin northwards, creating some of the most remarkable rocky shores on the

Rocky Shores

The Burren coast near Doolin has many superb examples of wave-cut limestone platforms, here covered in Thong Weed.

west coast through to Galway. So distinctive are they that Lewis includes detailed interpretive drawings of these shores in his masterwork.

Impressive wave-cut platforms of limestone can be seen at Fisher Street, the nearest shore to Doolin. From here, looking northwest, you can see the Aran Islands. Inhabited for millennia, these three monoliths rise vertically out of the sea as naked, giant slabs of karst scenery with not a tree to be seen. A visit to the largest island, Inishmore, across the rough Galway Bay, assails the visitor with a surprise of kilometres of high dry-stone walling, each enclosing small plots or fields, some of which are just metres across. Even more surprising is the lack of any gateways; to move animals a section of walling is removed and then rebuilt. Our first visit in 1975 revealed these 'fields' to have layers of seaweed in the soil. Originally, there was no natural soil on the islands except for what little developed within the cracks or grykes in the limestone. The soil now present has been painstakingly formed and developed by the inhabitants over many hundreds of years. Crushed rock, sand and seaweed are the essential ingredients, mixed with what little could be recovered from the grykes. In 1934 Robert Flaherty made the fictional documentary film *Man of Aran*, aiming to capture the culture of the people who were so dependent on the sea and shore, before their way of life was lost. It is now widely considered to be a factually incorrect and

A fascination for the shore

Dun Aonghasa, a prehistoric fort on the coast of Inishmore, Aran Islands.

over-romanticised depiction of life on the islands, but it is still well worth a watch for the atmospheric landscape. Basking Sharks *Cetorhinus maximus* cruise through the waters in summer, and they were hunted, but not with harpoons as shown in the film. Likewise, it shows the men smashing stones while women collect seaweed off the shore, when they actually each did both.

An unassuming spectacled man followed Flaherty as he filmed the shark-hunting sequence. This was Thomas Mason, an optician from Dublin, considered Ireland's finest natural history photographer of the time. He made a significantly better attempt to capture the cultural life of the west coast, and his account can be found in the evocative book *The Islands of Ireland*, published in 1938, embellished with his collection of monochrome photographs. Mason revealed the elements and the importance of the rocky shore in Irish culture, that it was not just a fascination but a crucial means of life support. He described the communities collecting a range of seaweeds for different uses. 'Dulisk' (probably Dulse *Palmaria palmata*) was picked by women and sent to Dublin for distribution around the country to be sold as a cure for hangovers. Meanwhile, men used metal hoops fixed in wooden handles to cut lower-shore wracks and kelp. This harvest, supplemented with storm-ripped seaweed from the strandline, would be placed in large baskets or creels to be carried

Seaweed from the shore being carried in the west of Ireland in the 1920s by women with goat skins protecting their backs (top), and by a donkey (bottom).

by women and girls with their backs protected by goatskins. Some seaweed was destined for the fields as fertiliser, while the rest went to construct large stacks above the shore. The tops were covered and protected from the elements by rushes so that the seaweed could dry. During summer months these ricks of seaweed were burnt to provide an ash from which iodine could be extracted on the mainland by chemical companies, ultimately for shipping across the world. Mason's book has become a classic and was reprinted 30 years later, sadly without the remarkable photographs.

The coast from Galway becomes increasingly rugged to the north around the counties of Mayo and Sligo, with both exposed and sheltered rocky shores. There are many of the latter around Clew Bay near Westport, which has many hundreds of islands. These are drumlins, egg-shaped hills formed of glacial material from the last Ice Age that became flooded as sea levels rose. The beautiful rocky shoreline continues round Donegal, back to Antrim and the Giant's Causeway. While there are some impressive islands to visit along the coast, Rathlin stands out as the example of a community that has embraced seaweed collection, not for fertiliser, but to eat. The company Islander Kelp, based on Rathlin, cultivates Oarweed, Sugar Kelp *Saccharina latissima* and Dabberlocks *Alaria esculenta*, growing these in nurseries within the Marine Conservation Area off the island's coastline. Seaweed products are sold locally or online; try some of their organic noodle cut, tagliatelle cut, or minced and whole-leaf organic fare. Ireland of course has a long-standing tradition of seaweed use and consumption. One notable species, *Chondrus crispus*, is a red seaweed with the common name of Irish Moss or Carragheen, and has been used for centuries to make a tasty sweet called Carragheen Moss Blancmange.

Further into the Atlantic lies the most remote rocky shore of the British Isles. The summit of the eroded core of an extinct volcano, Rockall has an approximate surface area of 570m^2 and stands on a submerged plateau called the Rockall Bank. Rockall is 369km west of North Uist, and between them lies the deep Rockall Trough or Channel. The elevated seabed and the possibility of the bank being a lost land has fuelled the myth that this was Atlantis. Today, with no freshwater available, it is home to a limited diversity of exposed rocky-shore invertebrates as well as a small number of Fulmars *Fulmarus glacialis*, Gannets, Kittiwakes and Guillemots.

From the northeast to the Humber

Heading back to the Scottish shore, the Moray Firth and the northeast of Scotland is a more gentle region, with harbours and inlets of sand and mud as well as rocky headlands and shores. Dolphins are the famous attraction in the Moray Firth, and they even have their own website (www.moraydolphins.co.uk). Good views can be had from the shores at Chanonry Point round to Troup Head. The latter is an RSPB reserve with Scotland's only mainland Gannet colony. Puffins and other auks can be seen here and at various other places down the coast where the cliffs are suitable above the rocky shore. Just south of Stonehaven, with its series of rocky coves and varying cliff heights,

Sea cliffs, home of seabird colonies in the breeding season, at Fowlsheugh, a coastal nature reserve in Kincardineshire.

are the easily accessible seabird cliffs at Fowlsheugh (also an RSPB reserve) with auks, Kittiwake and Fulmar. Bird numbers reach the dizzy heights of 130,000 individuals, all on the mainland. A little to the north, Dunnottar Castle on the cliff top creates an impressive backdrop to the rocky shores.

Rocky shores then pepper the coast southwards through the Northumberland Coast AONB down to the Durham Heritage Coast. Most sweeping bays are of large shingle and boulders, but good rocky shores exist in pockets, such as those at Seahouses on the way to the seabird strongholds of the Farne Islands. The explorer James Cook lived in Staithes on the north Yorkshire coast as a teenager, and it was here that his love of the sea started. The principal rocky-shore areas on this coast are at Runswick Bay and south to Robin Hood's Bay, before you reach Flamborough Head.

Professor Walter Garstang, who had been an assistant to the director at the MBA Plymouth Laboratory, became head of zoology at Leeds University in 1907 and opened their Robin Hood's Bay Marine Laboratory just south of Whitby in 1912, and Dr Jack Lewis (author of *The Ecology of Rocky Shores*) became director of the lab in the late 1960s. Plankton and fisheries studies were just some of the scientific work carried out here. As with so many of these establishments,

Robin Hood's Bay, Yorkshire.

research ended (1982) and the site has since become an interpretation centre for the National Trust.

Descending the cliffs at Flamborough Head, south of Robin Hood's Bay, we reach the last rocky shore of the east coast, an interesting chalky stretch with rock pools and a good range of seaweeds. This brings our circuit of the British Isles to an end. With a few exceptions such as Beachy Head, a search for the next rocky shore on a clockwise navigation would bring us back to Bembridge on the Isle of Wight, where our journey began.

* * *

Geographical position plays an important role in the natural history of the rocky shore, but the next chapter delves more deeply into the physical and chemical factors that influence this distinctive environment.

Patterns and zones

chapter two

Some years ago, on a clear April day, we were investigating a sheltered rocky shore when, within seconds, a northeasterly gale blew up. Such was its ferocity that the next day we were greeted by shopping-trolley-sized boulders that had been moved metres up the shore and positioned upside down. It was clear that these boulders were in the wrong place, for they had the wrong community of plants and animals for that height on the shore. The environmental conditions on the top of a boulder are also so different to those on the underside that they support completely different groups of organisms.

Vertical movement

Height up the shore was the key factor controlling what should have been on those displaced boulders, because that is what dictates how often communities are submerged in seawater. Tidal forces move water vertically up and down the shore; it may look as if the tide is moving in and out, but that is merely a consequence of slope.

The movement of seawater is of course mainly caused by the moon and sun, with the moon having the greatest effect as it is so much closer. Most places in the British Isles will experience two high and two low tides a day. The moon imposes a gravitational pull on our planet's oceans and so creates this daily pattern as the earth spins on its axis and a location's position changes, relative to it. Over the course of a month, when the moon and sun are approximately in line, their tide-raising forces combine to give spring tides. At this point, high waters are higher and low waters are lower than average. These large tides occur when the moon is full or new, which is at fortnightly intervals. 'Spring' has no seasonal context here, as it is from an old Anglo-Saxon word, *springam*, meaning to rise. In the intervening weeks, when the

OPPOSITE PAGE:
Zonation on a southerly shore of Skomer Island. At the top of the cliff Red Fescue marks a transition to more terrestrial conditions.

moon and sun are approximately at right angles, they oppose each other and as a consequence high tides are lower and low tides are higher. These smaller tides are referred to as neap tides (the etymology of which remains obscure). A third tidal pattern occurs over the course of a year. In March and September, at the equinoxes, the spring tides are greatest and the neap tides smallest. In June and December, at the solstices, the opposite is true. (Those interested in more details about tides and related matters can visit www.honeyhookphotography.co.uk/tides.html and follow the links.)

Knowing the state of the tide before visiting the rocky shore is essential. If you want to visit the lower shore, you need to coincide with low water of spring tides.

Tides and waves are just two of many environmental factors that determine which organisms live on the rocky shore, and where and how they live. The tide is particularly important in this respect, because as a result of tidal movements a significant environmental gradient exists on the seashore – sometimes over just a few metres – between fully marine conditions at the base and terrestrial conditions at the top. This gradient of conditions results in layers of animal and plant life, with significantly different species living at varying heights up the shore. The creatures that live here are anything but randomly distributed.

It is easy to take such information for granted. A great many people in the British Isles have reasonably easy access to rocky shores, and because of this we often underestimate how extraordinary they are. Everyone in the British Isles is within 113km of a habitat with some of the most varied and extreme conditions imaginable.

Biological zones

When the tide is out, rocky shores show horizontal bands of different colour, and these are referred to as zones. They may vary in width and in colour but they are almost always present.

The colours reflect the dominant or most apparent organisms within the zones; hence they are called **biological zones**. For example, some salt-tolerant vascular plants, namely Red Fescue *Festuca rubra*, Thrift *Armeria maritima*, Sea Plantain *Plantago maritima*, Sea Aster *Aster tripolium* and Buck's-horn Plantain *Plantago coronopus* occupy a zone above the true shore where salt spray affects the vegetation, but immersion never occurs. The colour of the high-level orange zone is usually due to the presence of orange and yellow lichens of the

Patterns and zones

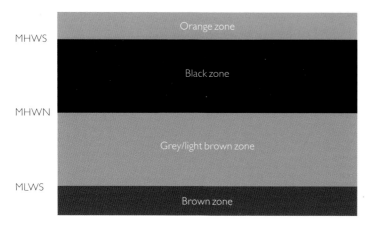

Basic rocky shore zones. The orange zone above Mean High Water of Spring (MHWS) tides represents a band of *Caloplaca* lichens. Below this, black tar lichens predominate. The grey/light brown zone, between Mean High Water of Neap (MHWN) and Mean Low Water of Spring (MLWS) tides, is a result of limpet and barnacle domination. At the base of the shore the brown zone shows where kelps dominate.

genera *Caloplaca* and *Xanthoria*, respectively. Below this exists a black zone dominated by black tar lichens of the *Verrucaria* genus. Further down, a grey or light brown zone indicates an abundance of barnacles and limpets anchored to the rock. At the base of the intertidal is a dark brown zone signifying species of large brown seaweeds, collectively called kelps. Depending on local conditions, some zones may not be obvious. The middle grey band may be obscured – or even absent – in sheltered conditions where the brown seaweeds Bladder Wrack *Fucus vesiculosus* and Egg Wrack may dominate instead.

Tidal heights and physical zones

Rocky shores can also be divided up into **physical zones** based on tidal heights, namely the upper, middle and lower shore. These tidal heights, and the zones derived from them, represent a useful way to divide up a shore for practical reasons.

Shore heights are measured in metres above a local reference point called Chart Datum (CD). This may be thought of as the lowest low tide in an area, referred to as zero metres, with all other shore heights being estimated relative to this zero point. In Milford Haven, for example, Extreme High Water of Spring Tides (EHWS) is approximately 7.8m above zero, Mean High Water of Neap Tides (MHWN) is 5.2m above zero, Mean Low Water of Neap Tides (MLWN) is 2.5m above zero and Extreme Low Water of Spring Tides (ELWS) is effectively Chart Datum or zero metres. These particular tidal boundaries have been defined because they are often used to divide the rocky shore into physical zones. Chart Datum is also used as a reference point on Admiralty charts; depth contours are measured below CD. Land maps

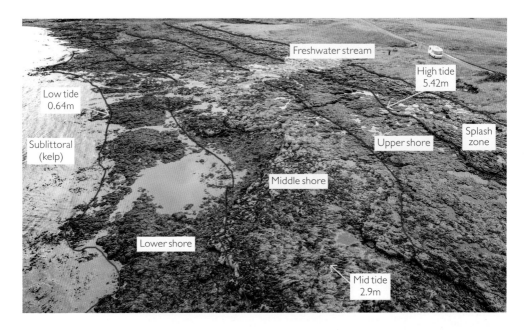

Shore zones on a typical sheltered rocky shore.

give heights above mean sea level, calculated from observations taken at Newlyn in Cornwall, and referred to as Ordnance Datum (OD). All local CDs around Britain are at known heights below OD; Milford Haven's CD is 3.71m below, for example.

Consultation of local tide tables will determine the relevant numbers for your own part of the country as they vary with the local tidal range. Variation may be considerable; the table opposite shows a range of values found within the British Isles.

The area between the average high and low waters of neap tides is called the middle shore. This 'true shore' is almost always covered and uncovered by the tide twice every day. The area above the lowest low tide and below that of the average neap low tide is called the lower shore. This is an area only uncovered during the low waters of spring tides. By contrast, the upper shore is that part above mean high water neaps but below the extreme high water of spring tides, which is only covered during the high-water period of spring tides. Above the upper shore there is a zone that is never immersed in seawater but is regularly splashed by it. Not surprisingly, this is called the splash zone.

Clearly, then, a significant environmental gradient exists on the shore. The further up the shore organisms live, the less time they spend under seawater. Since, in evolutionary terms, most rocky-shore organisms colonised the habitat from the sea, this imposes an environmental stress upon them that they must either tolerate or avoid.

Patterns and zones

Tidal range for selected sites in the British Isles

	Mean tidal range for spring tides (metres)	Mean tidal range for neap tides (metres)	Typical time for low water of spring tide, in hours a.m. or p.m. (GMT)
Straits of Jura	1.1	0.5	3–6
Portland Bill	2.7	0.9	1–4
Cork	4.7	2.5	11–2
Wick	5.8	1.7	4–7
Dover	7.2	4.1	6–9
Milford Haven	7.8	3.5	12–3
Bristol Channel	15.7	8.1	2–5

Source: UK Hydrographic Office (UKHO).

High (top) and low (bottom) spring tide at Jetty Beach, near Dale Fort. These photographs were taken on a day when the tidal range was at its maximum for this area, 7.8m.

Problems for rocky-shore inhabitants

The distribution of organisms on the shore is rarely determined by a single factor, but usually by a variety of interacting ones that vary with height. A good example is loss of water, which is a non-living (abiotic) factor. The abiotic components of the environment that an organism experiences vary more quickly and over a wider range of values in the air than under the sea. The higher up the shore an organism lives, the greater the length of time it will be exposed to these variations. Not only are the conditions more changeable, but also creatures are exposed to them for longer.

Abiotic factors

Water loss

On the rocky shore, water loss is the main factor that causes stress to the organisms that live there, since it can rapidly lead to desiccation. The basic physiology of most marine organisms is very simple: the concentration of their internal fluids is the same as that of seawater. Water may pass freely in and out so that there is no requirement for water conservation. When exposed to the air, however (an inevitable consequence of living on the shore), such organisms lose water by evaporation and may eventually die from dehydration. Survival on the shore therefore requires either the tolerance or the avoidance of water loss. For animals breathing by means of gills, this is a real problem.

Channel Wrack in a dehydrated (left) and moist (right) state. The image on the left shows swollen reproductive bodies at the end of some of the fronds.

The further up the shore, the greater the period of emersion, that is the time out of water, and the greater the potential for water loss. High-shore inhabitants require special adaptations; the upper-shore brown seaweed, Channel Wrack *Pelvetia canaliculata*, for example, is a master at tolerating desiccation. It can lose 95 per cent of its water content and survive (human beings would be in intensive care if they lost 10 per cent). Indeed, even from a black, brittle dehydrated state, Channel Wrack can resume photosynthesis, at the full rate, within 20 minutes of submersion.

Instead of tolerating water loss, many rocky-shore creatures such as periwinkles, topshells, limpets, mussels and Dogwhelks are able to avoid it, within limits, by having a waterproof shell. Others survive by living in rock pools or other microhabitats such as crevices and gullies, or under brown seaweeds. Sponges are especially prone to water loss, and their favoured habitats include crevices and the undersides of boulders. Of course, the most effective survival strategy is to reside at the bottom of the shore where desiccation is far less of a problem.

The risk of dehydration varies not only between zones but also within zones. In the upper shore, for example, the top edge is only

A shady microhabitat rich with sponges.

Creeping Chain Weed on a north-facing rock, accompanied by barnacles and rough periwinkles.

under water for 1 per cent of the time whereas the bottom of the zone is immersed for 20 per cent. The middle-shore immersion figures are between 20 and 80 per cent, the lower shore 80–99 per cent.

The direction a shore faces (its 'aspect') also strongly influences what may grow there, as sunlight and associated stresses will vary considerably. Red seaweeds in general are highly sensitive to sunlight, as it dries them out and destroys their pigments. Indeed, they may only survive on upper shores that face north. Creeping Chain Weed *Catenella caespitosa* is a moss-like red seaweed found almost exclusively on north-facing aspects; otherwise it is restricted to crevices and rocks under a canopy of Channel Wrack.

Temperature fluctuations

The temperature of British seawater varies little between day and night (perhaps 2°C in the southwest), or between summer and winter (perhaps 10°C in a similar area). Air temperatures on bare rock fluctuate much more wildly and quickly. In the upper shore, more than a 40°C seasonal variation is possible, from −12°C to +30°C.

Here, Channel Wrack is covered in snow, but at the same site in summer a thermometer placed in bright sunlight registered a maximum of 52°C.

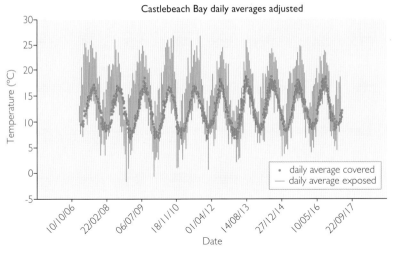

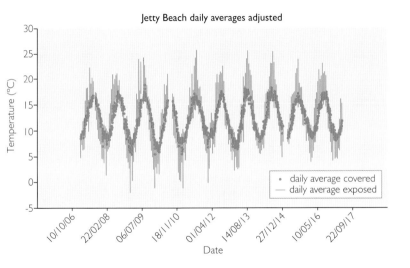

Data-logger data for a south-facing (top) and a north-facing (bottom) shore close to Dale Fort Field Centre. It is apparent from both shores' data how mild the 2013/14 winter was and how much colder the preceding ones had been. Orange: logger out of water; blue: logger submerged.

As with dehydration, the further up the shore that organisms live, the greater the temperature fluctuations they will be subjected to.

We collected temperature data at Dale Fort Field Centre from the two neighbouring shores for a number of years using data loggers fixed to the rock at Castlebeach Bay, a south-facing shore, and Jetty Beach, a north-facing one (see graphs above). Even though these shores are within a few hundred metres of each other, their differing aspects cause some interesting temperature differences. When the data logger was out of water the temperature frequently spiked above 20°C on Castlebeach, while spikes in this range were rare on the Jetty shore. The latter, facing away from the sun, had many more sub-zero values

Male limpet spawning in a rock pool in October 2010, Castlebeach, Pembrokeshire.

in winter than the south-facing shore. Temperature fluctuations when the logger was emersed (air temperature) were over a much wider range than when the logger was submerged, and this, unsurprisingly, was true for both shores.

Limpets are particularly sensitive to temperature fluctuations when it comes to reproduction. Though there are variations between species, spawning generally occurs when sea temperature drops to 11°C during the autumn, preferably aided by rougher waters (Lewis & Bowman 1975). These two abiotic factors help to synchronise the event so that the population on a given beach releases gametes en masse.

Variation in salinity

Upper-shore organisms are at greatest risk from potential fluctuations in salinity, and once again this is linked to emersion. If you are confident about the water quality, you could test this for yourself by tasting samples from small pools at different heights on the shore. Worldwide, the normal salinity of seawater is 3.4 per cent. This is often written as 34‰, which means 34 parts of salt per thousand of water. Around the British Isles it is nearer 35‰. However, this is a huge simplification, as there are at least 50 elements in seawater and many of them might be described as 'salts' in the strict chemical sense. The Red Sea is closer to 42‰, as it is almost a separate body of water subject to intense sunlight and hence evaporation. The shallow Baltic Sea can reach salinities as low as 15‰ owing to the abundance of freshwater running off the surrounding land, and surface salinities can be even lower where the less dense freshwater floats above the saltwater. Salinity can also vary greatly in tide pools with rainfall and sunshine. In the summer of 2006 we recorded an upper-shore rock pool in Pembrokeshire that reached 90‰; predictably there was nothing alive in it as living cells fall apart at salinities in excess of 60‰. The same rock pool in winter, after rain, registered a salinity of 5‰. While this may be an extreme example, it does demonstrate the challenges presented by both rain and sun on the shore.

Since marine organisms are usually in osmotic balance with their surroundings, if exposed to fresh or brackish water they will tend to absorb water by osmosis and swell up. For more primitive organisms, removing the excess water can result in the loss of important mineral ions. If the surrounding seawater becomes more concentrated, as it may well do in a rock pool on a sunny day, then water is lost from the creature's body.

The internal salinity of humans is maintained around 10‰; if water and salts are added or taken away our kidneys compensate accordingly. By contrast, invertebrate animals can struggle with variations in salinity. Sea anemones in a rock pool exposed to heavy rain that dilutes the seawater begin to take in the water and slowly swell until the tissues explode, causing the death and disintegration of the animals. Clearly they have no control over their internal fluids and are intolerant of change. These types of organisms are described as being stenohaline. By contrast, euryhaline species are able to live in environments that experience great changes in salinity. They can follow the two survival strategies that we have already encountered when considering water loss: tolerance and avoidance.

Animals that have the ability to tolerate changes in the salinity of their tissues are called osmoconformers. Common Mussels *Mytilus edulis*, when exposed to heavy rain, may absorb fresh water into their tissues but they are able to tolerate this until the tide returns. The acorn barnacles *Semibalanus balanoides* are also osmoconformers but

Rocky Shores

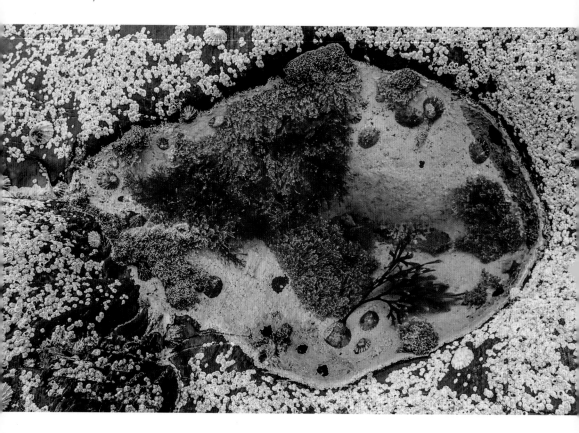

A rock pool surrounded by barnacles; in general, very few barnacles live under water in pools.

interestingly are very rarely found in tide pools (as can be seen in the photograph above). Perhaps the changes here are outside their tolerance range, but predation by prawns and shrimps may also be a significant factor.

Osmoregulators, however, can avoid changes in tissue salinity and are common denizens of rock pools and other areas of the shore where salinity might vary. One of the best examples is the Common Shore Crab, which has the ability to regulate movement of ions in and out of its body. Antennal glands, small bladder-like swellings at the base of the antennae, are used to osmoregulate. It is the only crab in Europe that can survive in salinities down to 4‰, and this helps to explain its abundance both on rocky shores and far up estuaries.

Light

The sun's influence on fair-skinned sunbathers is a fairly good illustration of the dangers of sunlight on rocky-shore creatures at low tide. Harmful effects come from a combination of radiation, heating and drying. It is often difficult to dissociate their separate influences, but this may be rather academic for the organism concerned if it has already expired. On its own, a single factor such as light may not be lethal, but combine several stress factors together and death results. Perhaps the most obvious adaptation to enable creatures to survive this harsh onslaught is a shell – and it is no coincidence that upper-shore animals are invariably protected by one.

At high tide, there is a shortage of light for the seaweeds below. Why might this be so? Light does not travel through water as easily as it does through the atmosphere. Firstly, the sun is never overhead in the British Isles (50–60°N), so its rays always strike the surface at an oblique angle. Sunlight reflecting off surface waters is consequently not passing through them for algal photosynthesis. Early in the morning, late in the evening, and in winter, very little light will penetrate the surface. Secondly, the sea is rarely crystal clear. Coastal waters carry a considerable load of silt, along with all the human flotsam and jetsam, and are usually turbid as a result of water movements in the shallows; plankton may also be abundant. All of these things physically block the passage of light. Thirdly, the wavelengths at the red end of the spectrum are selectively absorbed by seawater. Little red light – the wavelengths most useful for photosynthesis – penetrates more than 5m into the sea. In coastal waters blue light is absorbed and reflected by various particles, so green light penetrates deepest.

Wave action

Large waves can easily remove rocky-shore organisms and carry them up or down the shore to levels unsuitable for their survival; in so doing they can also have a profound effect on zonation patterns. Waves are a phenomenon of the sea's surface, and their effects are mainly felt where they hit the land, although of course spray is driven up well above that level and the undertow scours below it.

Because the middle shore receives the line of breakers four times on almost every day of the year, this is the zone that experiences wave action most frequently and where its effects are most pronounced. The largest waves may well break on the upper shore, but on a less frequent basis. Waves can also strike the shore with considerable force: 30 tonnes per square metre has been recorded. John's father, Anthony Archer-Thomson, was an officer on board the ocean liner RMS *Queen Elizabeth* in the 1950s, and on one occasion he saw an

Rocky Shores

Linney Head, Pembrokeshire, has deep water close by and is open to the Atlantic Ocean – these features conspire to produce enormous, powerful waves when conditions are right.

Atlantic breaker drench the foredeck and pluck a massive metal capstan from the deck in the process.

These comments apply to all shores, but the amount of wave action that a shore receives varies enormously. A shore that receives a lot of wave action is described as being exposed; one that does not is said to be sheltered. Students often look at us suspiciously when told that they are on a sheltered shore while being lashed by horizontal rain in a gale-force wind. Regardless of the weather, sheltered shores rarely, if ever, receive large waves; exposed shores do. The situation is complex but, as a guideline, a shore's exposure to wave action depends on four features: fetch, aspect, position and the depth of water offshore.

Fetch is arguably the most important factor determining an area's exposure. It is the extent or distance of open water that wind and waves can travel over before they hit the shore. If you have the opportunity, observe a large puddle or small pond the next time there is a strong wind blowing. At the windward end of the water, ripples or waves are very small, while to leeward they are significantly bigger. Extrapolate these principles to a really big pond, like the Atlantic Ocean. The waves that smash into the westerly coasts of Britain or Ireland can

Patterns and zones

Huge waves are not exclusive to headlands: the full force of an easterly storm in March 2018 swept away the shore, road and infrastructure, cutting off the east-facing village of Torcross, South Devon.

be truly enormous. There is open ocean all the way to Bermuda, 5,000km to the southwest. The greater a shore's fetch, the larger the waves that can strike it.

As mentioned previously, aspect is the direction the shore faces. On our western coasts, most of the wind and weather tends to come from the southwest, and so we refer to the prevailing wind as coming from the southwest. It is not surprising therefore that most exposed shores are open to the southwest in these parts of the British Isles.

The importance of position relates to the layout of the coastal features. For example, headlands are likely to receive the greatest forces from breaking waves. Exposure invariably decreases from a headland through to the relative shelter of a bay or inlet. Waves lose a lot of their power as they diverge and spread out within bays and into the shallows.

The depth of water offshore determines when the waves break, as they only do so when they reach shallow water. At a depth equal to half their wavelength (the distance between the crests) the crest steepens, and at a depth of one-third of the wavelength the wave breaks. So while large waves may break against harbour walls and vertical cliffs, on low-angle beaches they break well offshore and only wavelets reach

Rocky Shores

Bladderless Bladder Wrack on an exposed shore (left) compared with a sheltered form with many closely packed vesicles (right).

the land. It is the crest of the wave that does most damage to shore organisms, where energy from the wave is transformed into a linear punch that strikes the rock. Below the crest there is a zone where drag and scour are the predominant forces.

Wave action not only alters the composition of communities but can also strongly alter the appearance of species. The variation in the shape of limpets and Dogwhelks, depending on whether they live on exposed or sheltered shores, is discussed later on (in Chapter 7). Seaweeds also vary, and a notable example is that of the variants existing within Bladder Wrack populations. Air bladders engender an increase in the surface area of the seaweed's frond. The more bladders present, the greater the surface area. While bladders enable the frond to maximise absorption of light near the surface, a large area is more likely to cause the seaweed to be torn from the rocks when there is strong wave action. For this adaptable species the number of bladders is a good indication of exposure: the greater the wave action, the fewer the bladders. Bladders are also smaller on wave-exposed shores. Under extreme shelter in sea lochs the fronds are densely populated with bladders, giving them the appearance of bubble wrap, but at the exposed end of the spectrum a bladderless form, *Fucus vesiculosus* var. *linearis*, exists.

Geology and topography

The effect of the exposure of a shore to wave action may be modified significantly by its topography. Slopes, platforms, topographically complex areas and landward faces of rocky masses can provide sheltered areas on an otherwise exposed area of coast. Similarly, suitable microhabitats may provide refuges from the effects of sun and wind on all shores. Steep slopes increase the effective wave exposure of any type of shore. Slope will also affect drainage rates and, coupled with aspect, dehydration risk and temperature stress. Topography is often controlled, or at the very least modified, by the geology. Sedimentary rocks may, with one geological history, give flat platforms or gentle slopes. With another geological past they may yield vertical strata. The limestone of south Pembrokeshire is similar in composition to that of the Burren in County Clare, but the Irish wave-cut platforms and associated rocky-shore communities are largely absent from the steeply sloping sections of the Pembrokeshire coast. Even within Pembrokeshire's limestone topography and accompanying communities, there is considerable variation.

Once the geology has determined the topography of the shore, other influences tend to control community dynamics. Providing the rock is stable and not friable or eroding, then barnacles, limpets, mussels, snails and algal populations can all succeed, and variations in abundance – which might have had geological causes – are often overshadowed by the great variation imposed by other factors, many

The Burren, County Clare: a limestone wave-cut platform, with rock pools of many shapes and sizes.

of which have been mentioned above. Softer material such as chalk, shale and some limestones and sandstones can be unsuitable habitat for, say, barnacles, brown seaweeds and black tar lichen. Here, it may be limpets and more ephemeral green algae that succeed. Soft rock areas may be enriched by the presence of rock-boring animals. For example, the Common Piddock seems to favour the shale rock of the Jurassic Coast in Dorset. The Wrinkled Rock-borer *Hiatella arctica* does well in the limestone of the Gower. Both are bivalves adapted to live inside the rock strata.

As well as affecting community composition, rock type may also affect the behaviour of some of the species present. For example, limpet habits are quite different on rocks of differing hardness. On hard material limpets secrete new shell at their margins to complement the contours of the rock so that a good seal may be produced for water retention and to deter predators. On soft rocks limpets grind their shells around, in a circular direction, until the rock is worn away to fit the shell.

In most circumstances, however, the type of rock is less important than its profile and the topography of the shore.

pH

The pH of seawater varies between 7.6 and 8.4; in other words, seawater is slightly alkaline. Large pH fluctuations rarely occur in open water, and in this respect the sea acts as a buffer. Unfortunately, recent human activity and its effect on the global carbon budget has caused a worrying change to this comfortable status quo. Any major changes to ocean chemistry do not bode well for the survival of our species.

On the shore, the greatest variability in pH compared with the open ocean occurs in tide pools. Seaweed photosynthesis by day removes carbon dioxide and therefore boosts alkalinity. At night, seaweeds respire and expel carbon dioxide, which dissolves in water to create carbonic acid, reducing the pH. Animals respire all the time, of course, and animal excretion influences pool pH through the production of ammonia. Needless to say, the effects will vary with pool size and the relative amounts of animals and plants. Pollution can also have a strong influence: shortly after the oil spill from the *Sea Empress* in 1996 on south Pembrokeshire shores, a second 'oil spill' occurred when a container containing many hundreds of bottles of sunscreen was washed up, which contaminated local tide pools. pH skyrocketed for a few days until the contaminant dispersed.

Biotic factors

The challenges for rocky-shore organisms discussed so far relate to factors that are non-living: the physical and chemical abiota. These types of considerations are likely to be more significant higher up the shore, where creatures spend more time out of water. Lower down the shore, as immersion times increase, the abiotic environment for marine organisms (or rather those derived from fully marine forms) is more favourable. The downside to this is that there are more species and much greater abundance of those species, which means increased competition for resources and greater attraction for predators. These living parts of an organism's environment are called biotic factors.

As a general guide, the upper limit of the zone in which an organism can survive (its vertical range) is determined by abiotic factors. The lower limit of an organism's vertical range is often set by biotic factors.

Competition

Competition usually increases in intensity as you go down the shore towards the more favourable abiotic environment. If the environmental conditions are good then everyone wants to exploit them; for rocky-shore organisms, that mainly means competition for space, although seaweeds will also compete for light. Intraspecific competition (that between individuals of the same species) is potentially

Competition in action: limpets rearing up fighting for resources.

A close-up of several barnacle species, competing for space on the rocks.

more serious than interspecific (that between individuals of different species), because the competitors have nearly identical requirements. Competition for space is a direct form of competition and is sometimes referred to as 'interference' competition; the space is either occupied by an individual or not, there is no sharing. Competition for food, on the other hand, for example in neighbouring suspension-feeding colonies of sea-mats, sponges and sea-squirts, is called 'exploitative' competition. Here the resource is available to many and may be partitioned such that all benefit, though some more successfully than others. The importance of any competition in real-world communities is still a subject that engenders a great deal of debate. In harsh and/or disturbed environments, abiotic factors may act to decrease organism abundance long before competition plays a significant role. It is only in stable, benign environments that competition may play a major part in controlling community structure. This is more likely to be the case at the base of the shore.

In 1961, J. H. Connell, based at Millport, Scotland, published the results of a series of experiments on barnacles that have become classic examples of something called the 'competitive exclusion principle'. This states that where two species are after the same resource, the competitively superior species will out-compete the other and exclude it from the habitat. Connell demonstrated this with two barnacle types that have overlapping distributions in the British Isles. The acorn barnacle *Semibalanus balanoides* is a northern species living across Scandinavia and northern Europe as far south as Britain. Stellate barnacles of the genus *Chthamalus*, on the other hand, have a more southern distribution around the Mediterranean, reaching the British Isles at the northern end of their range. Both are similar in size and appearance and have an outright need for space on the rocks. The southern species can occupy most of the shore, including the upper shore with its associated high summer temperatures. The northern species rarely survives high up the shore, doing better in the cooler areas further down. However, where the two populations meet in the middle shore, the northern species out-competes the other for space on the rock. It does this because it grows faster and its calcareous plates, which increase in size as the animal grows, are stronger, crushing the softer ones of *Chthamalus*. So the *Chthamalus* population is competitively excluded by *Semibalanus* where they both could survive but *Chthamalus* persists where the abiotic environment is too harsh for *Semibalanus* – a good example of interspecific competition.

Such considerations also apply to rocky shore plants. Seaweeds such as mature Bladder Wrack prevent young sporelings from developing beneath them by shading out the light. They also 'sweep' the rock with their fronds, as they move about in the surf, which prevents young algae (and other organisms) becoming established. Only when the adult plants die will the youngsters be able to grow.

Competitive exchanges are usually, by their very nature, to the detriment of at least one of the parties involved, but such interactions need not necessarily be negative. Some big filter feeders, such as Common Mussels, can affect local water flow as they take in liquid, bringing more plankton and particulate-rich fluid to the local environment. In this way food supply may be enhanced such that it favours or 'facilitates' other organisms. Mussel beds may also retain more moisture at low tide than bare rock would and hence reduce the risk of dehydration for resident organisms. These adjustments to local conditions are such that Mussel beds may support a community of up to 100 other species as a result (Little *et al.* 2009). Where organisms modify a habitat to the benefit of others, they are sometimes called 'foundation species', 'ecosystem engineers' or 'habitat formers'.

Predation

Rocky-shore organisms are exposed to a double set of predators (see Chapter 11), which seems a trifle unfair. Marine predators such as crabs and fish abound when the tide is in, giving way to terrestrial ones, including birds and people, when the tide is out.

Predation rates can be impressive. A number of ecologists have studied shore birds including Rock Pipits *Anthus petrosus* in Cornwall over several decades and have found that these birds can spend up to eight hours feeding, eating up to 33 Small Periwinkles *Melarhaphe neritoides* a minute, or 14,000 a day (Feare & Summers, in Moore & Seed 1985). This is matched by the Turnstone *Arenaria interpres*, but they seem to avoid competition with the Rock Pipits by consuming larger-sized snails. Oystercatchers *Haematopus ostralegus* can pack away 300 Cockles *Cerastoderma edule* each day.

Modifications to the zonation patterns

Although practically all British and Irish rocky shores show some sort of zonation, the detailed pattern is almost infinitely variable. The main reasons for variation in zonation patterns from place to place are differences in tidal range and exposure to wave action, while various estuarine effects are important near the mouths of large rivers. Although you may have to travel large distances to observe a change in tidal range, most local variation is due to changes in wave exposure. So far we have discussed the idea of vertical zonation on the shore, but differences in exposure along a stretch of shore may be thought of as a kind of horizontal zonation. If you have access to a stretch of rocky coast that goes from a sheltered bay out to a headland, have a look (ideally through binoculars) for changes in community composition as exposure increases. You may also observe changes in zonation patterns.

The effect of tidal range

Where waves are bigger than the tidal range, exposed shores show little sign of zonation, except in the usually very extensive splash zone. The sharpest zonation patterns are seen on sheltered shores with tidal ranges of between 1 and 3m – a glance at the tide tables for such places will show that the predictions are almost the same for many successive days. Once the range exceeds 10m there is considerable

Horizontal and vertical zonation; the increase in exposure from right to left causes a decrease in seaweed cover and an increase in barnacle abundance.

blurring of the zonal boundaries, because tidal heights differ greatly from day to day. On such shores, the splash zone is comparatively restricted because little or no spray carries up from waves breaking on the middle or lower shores.

The effect of wave exposure

The bigger the waves, the greater the amount of spray that will be produced, and by spraying high up the shore, regions normally prone to drying out will be moistened. Biological zones extend higher up the rocks as exposure to wave action increases, even when the tidal heights remain exactly the same. This reinforces the idea that desiccation may be an important limiting factor at the upper limit of various organisms' vertical ranges.

Black tar lichen *Verrucaria* spp. on sheltered shores within Pembrokeshire's Daucleddau estuary may occupy a zone that is only a few centimetres deep. At St Ann's Head, to the north side of the entrance to the Milford Haven Waterway and open to the full force of Atlantic breakers, the black tar lichen zone may be over 50m deep. Lichens in the splash zone occupy a region where it is too salt-sprayed for terrestrial vegetation but not wet enough for

Rocky Shores

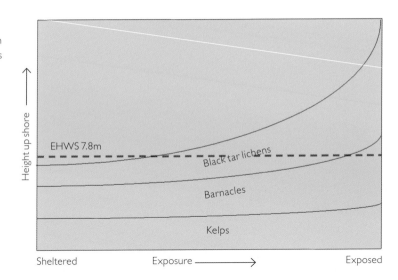

The effects of exposure on zonation. EHWS 7.8m (red dotted line) denotes the height of the predicted highest tide of the year for the Port of Milford Haven.

intertidal organisms to survive. This salty no-man's-land is small in vertical extent where the waves are small but increases in size with exposure to large, powerful waves.

Wave action can also extend the upper limit of where animals such as barnacles can survive, because the extra splash provides a supply of food and water above that which the tide would supply. In cases of extreme exposure, certain barnacles can survive in the splash zone. Many years ago, in Eshaness, Shetland, we found a rock pool containing gutweed and sea lettuce (green seaweeds), limpets, topshells and periwinkles. The pool was at least 50m up the cliff, well above the height of the highest tide, but was fed as waves crashed onto the shore below and water was channelled up a crevice in the cliff face, replenishing the pool on a regular basis.

But even exposed shores are subject to periods of calm. For organisms to survive above their average zonation maxima as a result of extra splash and spray, they must also be able to retain enough water to survive in between bouts of rough weather. The upper extension of the kelp zone under the influence of extra splash and spray is minimal, as kelp plants are extremely sensitive to water loss (they die if they lose more than 20 per cent of their water content). They also have a large surface area from which to lose water.

It is all very well to talk theoretically of a shore's exposure to wave action, another matter entirely to measure it. There are three approaches, all presenting difficulties. The first is to use a direct measurement of the wave action. People have tried measuring wave

A stormy day at Freshwater West, Pembrokeshire. These conditions present a real challenge when it comes to directly measuring exposure.

pressures using pressure transducers, or dynamometers; or they have tried to measure drag forces with drogues attached to spring balances. Others have measured the height of the waves against marks on a cliff or harbour wall. In principle, direct measurement is the obvious approach but, in practice, it is just not possible to obtain long enough runs of data to describe even one site. Even if it were, such data are unlikely to reflect the complex interactions controlling the organisms' habitat. Survival rates of such measuring instruments tend to be poor!

Indirect measurements of exposure may be inferred from published maps and charts by categorising the factors that control the size of waves. During the late 1950s, Bill Ballantine listed shores in Pembrokeshire that are open to the Atlantic over a 67-degree angle to fetches of more than 3,219km as 'extremely exposed', those open over angles of 40–50 degrees to fetches over 322km as 'very exposed', and so on down to shores open over less than 60 degrees with fetches of less than 16km, which he listed as 'very sheltered'.

A third and possibly most useful measurement of exposure is to use biological indicators – the organisms themselves – to quantify the long-term environmental conditions. A biological exposure scale devised by Bill Ballantine in 1961 does just this.

By their very survival, the biological communities on any rocky shore represent an integration of the complex variables comprising 'exposure'. While still a PhD student working at Dale Fort Field Centre, Ballantine noticed that the local rocky shores supported different species depending on their exposure to the Atlantic Ocean. As you might expect, certain organisms, such as barnacles, limpets and crustose lichens, are very good at coping with wave action; others, such as crabs, some periwinkles, topshells and many seaweeds, are not. Ballantine devised a numerical scale from 1 to 8 on which you can assess a shore's exposure to wave action by measuring the abundance of certain fairly common rocky-shore species.

Ballantine's Exposure Scale (BES), as it became known, can be used to give any rocky shore in the British Isles an exposure grade, depending on the characteristic pattern of communities that are present. This scale is a biologically defined measure of exposure, as it is the organisms themselves that allow a score to be assigned; no physical (abiotic) measurements are required. An initial or rough guide to a shore's exposure grade can be given with reference to a simple diagram (opposite). This shows a range of shores from left (grade 1, extremely exposed) to right (grade 8, extremely sheltered). Exposed shores are barnacle-dominated, sheltered are seaweed-dominated. The method is to stand facing the chosen shore, preferably at low water, making sure it is safe to do so. You then compare your shore with the diagram and see where it 'fits', then go to the number at the top of the diagram and read off your shore's exposure grade. The score is sensitive to slope; if your shore varies considerably you may assign different scores to different sections of the shore.

Biological exposure scales use the concept of indicator species: organisms characteristic of a particular range of environmental variables that may be used as indicators of those variables. Some lichen species, for example, are useful indicators of air quality; some freshwater macroinvertebrates are used to indicate the pollution status of a body of water. In the case of exposure, we use a cocktail of species, some characteristic of sheltered rocky shores, others of exposed rocky shores. Some marine biologists object to biological exposure scales on principle because they are based on a circular argument. A shore is exposed because it has a certain community of organisms, and it has this community because it is exposed.

Unfortunately, exposure is not the only factor influencing the distribution of the indicator species selected by Ballantine for his scale.

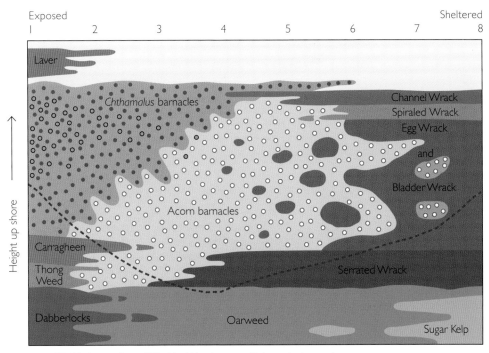

- ◉ Bladderless variant of Bladder Wrack
- --- Upper limit of encrusting red algae
- ● *Chthamalus* barnacles
- ○ Acorn barnacles

Ballantine's Exposure Scale diagram. Adapted from Figure 1 in Ballantine (1961).

Some are northern species, reaching their southern limit in southwest Britain; others are southerners that do not extend much further north. Ballantine himself regarded his scale as valid only within 160km of Milford Haven, and recommended that workers devise their own native scales based on local observations. Amended versions of the scale for northwest Scotland and Norway exist, and other versions have been produced for shores as far away as New Zealand. As long as one is aware of the shortcomings of these biological exposure scales, they can be extremely useful tools.

Some of the sites used by Ballantine to define his exposure scale did not key out for him at the same grade when he revisited them 20 years later. This may have been due to a change in physical exposure, but the distribution and abundance of the indicator species can and does vary in seasonal, short- and long-term cycles. Some monitoring work done in Milford Haven Waterway over a period of more than 30 years has yielded some fascinating results that give an insight into the question as to how much variation is 'normal' in rocky-shore populations. Details of this work are included in Chapter 13.

Rocky Shores

The Wick, Skomer Island, showing an 'ideal' rocky shore of uniform slope without the complications imposed by microhabitats.

Microhabitat availability

A 'classic' rocky shore might consist of solid rock at a 45-degree slope and contain no pools, crevices or other features to confuse the pattern enforced by the tidally induced environmental gradient. Such shores do exist – the Wick on Skomer is one such example – but they are rare.

Usually there are complications to alter and confuse the simple zonation patterns as a result of geological vicissitudes. Rock or tide pools provide such intricacies, and are a classic microhabitat beloved by generations of rocky-shore explorers. Over a tidal cycle, temperature, pH and salinity may fluctuate greatly, but water loss is rarely an issue unless the heat is so intense that the pool evaporates, leaving nothing but salt crystals.

The exigencies of rock-pool life make for fascinating, often unique, little ecosystems. Rock-pool specialist species exist and warrant attention, as we will see in the next chapter. Crevices, large and small, retain more moisture than open rock, and any occupants are likely to benefit from a greater amount of shade. Macro-algae (the larger seaweeds) can have a similar ameliorating influence, as well as providing living space and potentially food for a number of shore organisms. Crab species and other crustaceans such as the amphipods and isopods thrive in the moist, cool conditions provided here.

Some seaweed microhabitat relationships are highly specific. Egg Wrack can often be found with a specialist red seaweed, Wrack Siphon Weed *Vertebrata (Polysiphonia) lanosa*, growing on its surface. The nature of the relationship is obscure, but the red alga is known to penetrate its host's tissues rather than simply living on the surface. The reproductive structures of the Egg Wrack are located on stalks, and when it releases its gametes, the stalk drops off. Being a rather primitive organism, the shedding of these stalks produces sizeable gaps in the brown seaweed's epidermis for the red alga to invade. Is the red seaweed a parasite or just using the Egg Wrack as a substrate? It is unclear and still a mystery; after all, if it is a parasite why does it still have photosynthetic pigments? Certainly the red seaweed obtains a lift above the densely shaded area beneath the large brown wrack. Perhaps the red just gains a measure of protection from the sun's rays, and the specific host's zonation position is one at which the red seaweed can survive. Using labelled carbon 14 (radioactive carbon incorporated into sugar compounds can be used to trace nutrient pathways in chosen target species), it is possible to show that the transfer of sugars may go both ways between red and brown seaweed. The jury is still out on the exact nature of this relationship, but this example serves to remind

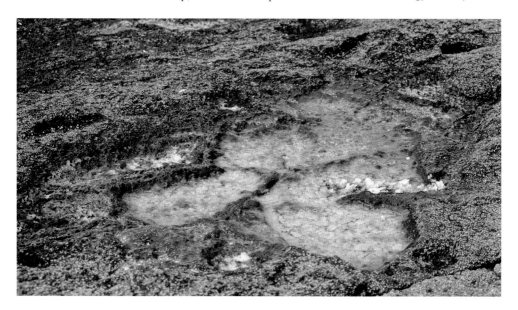

A rock pool showing extreme evaporation, leaving just salt crystals.

The brown seaweed Egg Wrack, showing its large egg-shaped gas bladders, plus its epiphytic red alga, Wrack Siphon Weed.

us how difficult it can be to gain a full understanding of the detailed associations that contribute to the rocky-shore ecosystem.

Other microhabitats, such as the shaded faces of overhangs and boulders, can support a plethora of sedentary species including sponges and sea-squirts. Such species also attach to crab carapaces, where they may be joined by barnacles. Even the surfaces of discarded shells, which are often used as temporary residencies by hermit crabs, make ideal, hard material for attachment. Any microhabitat can distort simple zonation patterns and make an analysis of distribution and abundance a much more complex – but rewarding – task.

A Common Shore Crab with bryozoans (white 'cellular' encrustations on the limbs) and parasites (*Sacculina carcini*, a yellow swollen mass, bottom middle of image) on its carapace.

Freshwater on the shore

Microhabitats may also take the form of streams and other sources of freshwater running over the shore, which can greatly influence zonation patterns.

Green seaweeds, *Ulva* species in particular, have a great tolerance for changes in salinity and often indicate a freshwater source by their presence. It was noticeable during the wet winter of 2012/13 that ephemeral freshwater streams appeared on numerous shores; some established streams changed their course dramatically. Creatures in the path of this sudden freshwater influx were doomed unless, like the Common Shore Crab, they could osmoregulate. Many molluscs died, as did some seaweed species that were less tolerant of osmotic stress. *Ulva* species thrived, partly because of their tolerance of reduced salinity but also as a result of reduced grazing pressure; many herbivores, like limpets, were not able to feed successfully in such a low-salinity environment.

A freshwater stream over the shore at Newgale, Pembrokeshire. Gutweed is doing particularly well in this variable-salinity environment.

Horned Wrack, Ullinish, Skye.

Ulva species are opportunistic; they grow quickly and can take over areas in a matter of days and so are ideally suited to exploiting temporary environments. Streams that are less transient and survive year on year, or estuary shallows, support those brown seaweeds that are specialised for fresh or brackish water. Horned Wrack *Fucus ceranoides* is a case in point, especially so in the north where permanent streams are more prevalent. In this way, brackish water on the shore can substantially alter the community zones.

The rocky shore at Watchet Harbour, Somerset, is near to the Leonard Willis Field Centre where Dr John Crothers lived and worked for almost three decades. He studied mollusc populations and made many scientific discoveries here, including that the survival rates of Common Mussels were higher within a stream running across the shore than on the rocks either side. The Mussels were able to close their shells and keep out freshwater during low tide, when salinity was appreciably reduced. Their main predator on that shore, the Dogwhelk, was unable to tolerate conditions in the stream and so the Mussels had a predator-free refuge for at least part of the tidal cycle, and hence their survival was enhanced (Wilson *et al.* 1983).

Tidal rapids

As the tide drops, seawater usually drains slowly from the shore, but where water passes through narrow constrictions, speeds of flow may increase dramatically. On a small scale this could be a large rock pool, which on emersion drains to form a small waterfall, flowing through a crevice. This increased flow may attract large numbers of sessile creatures such as sponges and sea-squirts, eager to make use of the replenishing current of nutrient-rich water. These localised areas of increased flow have knock-on effects on adjacent communities. On a large scale, the result is a dramatic set of rapids. As discussed in Chapter 1, these are common on the west coast of Scotland where flooded valleys and sea lochs empty to the sea through narrow channels, and similar conditions produce rapids in County Cork at Loch Hyne, where the marine station is based. In some sheltered areas of Milford Haven, Wales, and south Devon there are rapids on a smaller scale, which issue from tidal lagoons as they empty on the ebb tide.

Large tidal rapids are sought out by discerning white-water rafters looking for an exciting ride, but the small ones are particularly

Tidal rapids at Loch Obe, Barra, complete with some students from the 1970s.

Rocky Shores

The undersides of rocks from tidal rapids are encrusted with sponges, sea-squirts, sea-mats and tubeworms. All are suspension feeders.

interesting for the naturalist. Rates of flow are greatest during spring tides. Typically the greatest change in the intertidal communities will be on the lower shore. The community living here will be modified by conditions on the surrounding shore, which is most likely to be sheltered, around BES 7 and 8, with upper shores dominated by Channel Wrack and Spiraled Wrack *Fucus spiralis*. The lower shore, where the current speeds are greatest, will have species more typical of wave-exposed areas, such as the tougher kelps Oarweed and even Dabberlocks. In this way, again, our zonation patterns have been modified. Sessile, strongly attached animals are favoured, including sponges, sea-mats, sea-squirts, horse mussels and hydroids that feed on the near-constant stream of detritus passing over them. Some mobile species that make use of material suspended in the water, such as the echinoderms, feather-stars and brittle-stars, also thrive here. Tidal rapids, with their rich current-fed waters, are a mecca for species that can tolerate the conditions; consequently such areas are often high in biodiversity.

At low tide the outflow of water slows until it eventually stops altogether. There is a brief period of slack before the flow reverses, current speeds increase and water floods back in. Seaweeds subjected to this twice-daily reversal of flow can become weakened; damaged fronds and stipes are commonplace. During the ephemeral, quiet periods of slack water, mobile predators such as crabs and fish will move in to feed. Herons and other wading birds are a common sight too at this time.

A walk down the shore

In this chapter so far we have described the zones – biological and physical – that can be used to subdivide the shore, the factors – abiotic and biotic – that affect rocky shore organisms, and the way in which zonation patterns may be modified by other influences. To conclude, let's have a look at some typical organisms visitors might encounter on both sheltered and exposed shores. The simplest way to do this is to walk from the top of the splash zone down to the lowest point exposed at low tide. Choose a day with a good spring tide by checking the local tide tables. Wherever you decide to explore along the coastline, begin by scrabbling to the top of the splash zone.

It should be quite easy to find the extreme limit of this zone, as just below the first flowering plants (commonly species such as Red Fescue, Thrift and Sea Plantain) the rocks are typically dominated by lichens. This is a broad zone in exposed situations, narrow in sheltered. In fact in very sheltered locations you might even miss it, as it can be reduced to less than a hand-width. Within the splash zone, Sea Ivory *Ramalina siliquosa* is more common at the top, then there may be a selection of grey encrusting lichens including various species of *Lecanora* interspersed with the yellows of the *Xanthoria* genus and the creamy white Crab's-eye Lichen. To seaward, *Caloplaca* lichens thrive and mingle with black tar lichens of the *Verrucaria* genus, which come to dominate the top of the intertidal. Watch out for Sea Slaters *Ligia oceanica*, woodlouse-like creatures, scuttling about, and Bristletails *Petrobius maritimus* darting amongst the rock crevices. Rough periwinkles appear towards the bottom of the zone, and in

Splash zone, West Angle Bay, Pembrokeshire, showing many common coastal lichen species, including Sea Ivory (green), various *Xanthoria* (yellow) and *Caloplaca* (orange) species, *Lecanora* (grey) and Crab's-eye Lichen (white).

exposed areas they are joined by Small Periwinkles. The latter species is particularly abundant in crevices and dead barnacle 'shells', but you will need to look very carefully as they are just a few millimetres long.

A few lichens continue down into the upper shore, namely *Caloplaca* and black *Verrucaria* species that may thrive here in exposed areas. Green Tar Lichen *Verrucaria mucosa* appears towards the lower half of the zone; wet patches of this lichen are noticeably greener than their black up-shore relatives. As you move down, the sudden appearance of brown seaweeds marks the transition to the tidal zone. Look out for the upper-shore specialists Channel Wrack and Spiraled Wrack. In sheltered spots these weeds may dominate the zone; towards the more exposed end of the spectrum they will be absent. In suitable microhabitats such as crevices, under brown seaweeds and on north-facing rocks, the red seaweed Creeping Chain Weed often does well, sometimes accompanied by other moss-like reds depending on local conditions. Green seaweeds, especially some of the *Ulva* genus, are extremely tolerant of environmental variation and can be found all over the shore.

Rough periwinkles continue to thrive and animal diversity steadily increases as abiotic conditions improve, but this high up the shore shelled organisms, like limpets, *Chthamalus* barnacles, and Small and rough periwinkles, do best in exposed areas where spray keeps them moist. On exposed shores these will be very obvious owing to the lack of seaweed, but back on the sheltered shore at the same level you will need to hunt for creatures that are avoiding these dry conditions. Try

Typical upper-shore communities from an exposed shore, left (Dale Point, Pembrokeshire, BES grade 2), and a sheltered shore, right (Black Rock, Pembrokeshire, BES grade 7).

looking among seaweed and around pools for Toothed (Common) Topshells *Phorcus lineatus* and Purple Topshells *Steromphala umbilicalis*, which will be increasingly abundant (although absent from eastern coasts). Edible periwinkles *Littorina littorea* and flat periwinkles may be present at this height, depending on local conditions. Topshells and periwinkles are both snails with spiral shells, as opposed to the dome-shaped version of a limpet, and can be tricky to tell apart at first. Try picking a few up and looking at the shell opening. Topshells have mother-of-pearl around a perfectly circular operculum (trap-door), which is used to close the opening when the snail is in retreat. Periwinkles have no mother-of-pearl and their trap-doors are teardrop-shaped. The Beadlet Anemone *Actinia equina* is the one soft-bodied animal that may be common if suitable microhabitats, pools and crevices, are available.

An extensive search of the upper shore will not provide you with a very diverse community. This is still a harsh environment with very little seawater available. Creatures need to be specialised to survive the extreme conditions, but those that are will be able to live here with little competition except from individuals of their own species.

Keep walking down towards the falling tide and you will notice that biodiversity is rising as the abiotic conditions improve for marine organisms. Lichen species are reduced in numbers but Green Tar Lichen thrives, especially in areas that are not covered in seaweed. If you are on an exposed shore dominated by barnacles and limpets, navigation into the middle shore can be tricky, as zonation is not clear.

Typical middle-shore communities from an exposed shore, left (Dale Point, Pembrokeshire, BES grade 2), and a sheltered shore, right (Llawrenny, Pembrokeshire, BES grade 7).

However, cast your eyes about for dark patches standing out among the expanse of pale barnacles; these will be quite obvious from a distance of several metres. Close observation will reveal that these are dense mats of a brittle moss-like lichen, Black Lichen *Lichina pygmaea*. This lichen microhabitat is sometimes full of Coin Shells *Lasaea adansoni*, and pseudoscorpions *Neobisium maritimum*, but you will need to look closely.

Meanwhile, if you have been walking on a sheltered shore there will be no *Lichina* present, but a good indication you have reached the middle shore will be the typical brown seaweeds Bladder Wrack and Egg Wrack. Egg Wrack does best in extreme shelter, with Bladder Wrack occupying shores of sheltered through intermediate to exposed conditions, where it survives in its bladderless form (var. *linearis*). Red seaweeds become more numerous; a typical middle-shore representative might be Pepper Dulse *Osmundea pinnatifida*, which can survive in most exposures, albeit in a rather stunted form at the exposed edges of its range. The variety of animals also continues to increase as we head down the shore, as does the abundance of individuals within each species. Limpets, Dogwhelks, Purple and Toothed Topshells, rough, edible and flat periwinkles, barnacles, mussels and even chitons (primitive molluscs that look a little like legless woodlice) might all be present, depending on local geography and conditions. More soft-bodied creatures are also apparent; anemones and various species of sponge may be discovered, especially in any gullies and under overhangs.

Ideally you will have timed your excursion so that you reach the lower shore at low tide. Although this zone represents the abiotic optimum for many shore organisms, as we have seen, there is a price to pay for living here. Competition and predation become more of an issue, and those flora and fauna that cannot cope with these biotic pressures may only survive higher up the shore. You will no longer encounter lichens, for they cannot tolerate such prolonged periods of immersion and consequently do not live here. You will likely find the typical brown seaweed of the lower shore, Serrated Wrack *Fucus serratus*, especially towards the top and middle of the zone, with kelps of various species, depending on the exposure, towards the bottom. If you are in an area where wave action is particularly acute, Serrated Wrack suffers and is replaced by Thong Weed. Strangely, Thong Weed makes an appearance again in extreme shelter, as found in sea lochs off Scotland's west coast. This seaweed can cope with any degree of wave action but not competition with Serrated Wrack, so if

Typical lower-shore communities from an exposed shore, left (West Angle Bay, Pembrokeshire, BES grade variable), and a sheltered shore, right (below the MBA, Plymouth, BES grade 7).

you find the wrack you may not encounter Thong Weed. Reds abound especially under the canopy of the bigger, faster-growing browns and do well at the base of the kelp forest even into quite extreme exposure.

Be prepared to spend more time searching this zone than all of the others, as animals do well here, just like the seaweeds – but for safety's sake keep your eye on the state of the tide. There will be the rather more desiccation-sensitive examples such as sea-mats, sea-squirts, hydroids, crabs, fish, sea-slugs, and sea-hares that will all add to the excitement. The Grey Topshell *Steromphala cineraria* may be encountered on a spring low tide, as might the impressive Painted Topshell *Calliostoma zizyphinum*, while the Turban Topshell *Gibbula magus* is locally common on south and northwest coasts. Like many other temperature-sensitive species, it is absent from the east coast of Britain as this coast lacks the ameliorating influence of the Gulf Stream. The lowest tides may even reveal some echinoderm species such as the Common Starfish *Asterias rubens*, especially if Mussels are abundant, and the omnivorous Shore Sea Urchin *Psammechinus miliaris*, which may be found grazing amongst the kelp stipes.

* * *

To explore lower down the shore you will need a wet suit and snorkel. Alternatively you could lie out on a rock and peer into a tide pool, where the conditions will be different. To start with, desiccation does not seem to be an issue. Or is it? We need a new chapter.

Rock pools | chapter three

'There is nothing, now, where in our days there was so much. Then the rocks between tide and tide were submarine gardens of a beauty that seemed often to be fabulous, and was positively delusive, since, if we delicately lifted the weed-curtains of a windless pool, though we might for a moment see its sides and floor paved with living blossoms, ivory-white, rosy-red, and amethyst, yet all that panoply would melt away, furled into the hollow rock, if we so much as dropped a pebble in to disturb the magic dream'

Edmund Gosse, poet and literary critic, son of P. H. Gosse, from his autobiographical memoir *Father and Son* (1907)

Rock (or tide) pools are probably the reason why so many people love exploring seashores. Who can resist examining these beautiful green and pink aquatic gardens to see which animals might be discovered lurking within the seaweed fronds? The English naturalist Philip Henry Gosse (1810–1888), credited as the main inventor of the seawater aquarium, gained the necessary inspiration from his observations of rock pools on the Devon coast. Gosse launched the aquarium craze when he created and stocked the first public one in London Zoo in 1853, and in doing so encouraged a fashion that was to bring aquaria into many Victorian homes. His 1854 book *The Aquarium: an Unveiling of the Wonders of the Deep Sea* introduced the word 'aquarium' for the first time and described how readers could make their own. Later that year he produced a recipe for making seawater, as

OPPOSITE PAGE: Upper-shore rock pools, West Angle Bay, Pembrokeshire.

ABOVE: Hibberd's 1870 drawing of a Gosse aquarium.

Rocky Shores

Dr Robin Crump with Mark Burton and Anne Bunker, surveying Brooding Cushion-stars in a rock pool at West Angle Bay, Pembrokeshire.

he maintained that it was 'just too tedious to travel to the coast' to collect the real thing. His original recipe, although not perfect and subsequently improved, demonstrated that homemade seawater could sustain some crustaceans for two years.

Marine biologists' addiction to rock pools can inevitably result in them having favourites. Our friend Dr Robin Crump has visited his chosen pool on the shore of West Angle, Pembrokeshire, for the best part of 40 years. His monthly outings enable him to count numbers of Brooding Cushion-stars *Asterina phylactica*, which he studies. But we tend to forget that these microhabitats, manifest as discrete entities at low tide, are subsumed within the ocean as a whole when covered by seawater.

Rock-pool conditions

Rock pools are notable not only for the creatures they may contain but also for the way in which they modify zonation patterns. Many species of seaweed, in particular kelps and red seaweeds, which are typical of the lower shore, have their vertical ranges extended by the presence of suitable pools. Dehydration is not a problem within this microhabitat (unless conditions are so extreme that the water evaporates), and the water helps to protect seaweed pigments from strong sunlight. The living conditions of rock pools are more stable than those on the surrounding rock, although far less so than in the adjacent ocean. When initially emersed, pools start out with the same temperature, pH, salinity, oxygen and carbon dioxide values as the open water. But, within minutes, and until the tide returns, conditions begin to change, depending on the size of the pool, its aspect and shore height. Shallow, south-facing, upper-shore pools will alter the most as the water recedes; deep lower-shore pools will be the most stable and are effectively extensions of the sublittoral, with a suite of species to match.

Wide salinity and temperature fluctuations may be the most stressful for rock-pool inhabitants, and these variations are imposed on the pools externally. But in a pool packed with seaweeds, photosynthetic activity can reduce carbon dioxide levels, raise the pH and increase oxygen concentrations to over three times saturation values. Bubbles of oxygen can easily be seen if you observe a suitable pool for just a few minutes. Working with students over several decades, we have collected data from rock pools at both Kimmeridge, Dorset, and in Milford Haven, Pembrokeshire. In one pool a pH of over 9 was recorded during the day, but at night, when seaweed photosynthesis had ceased and all of the creatures were respiring, the pH fell to 6; carbon dioxide levels had increased and, perhaps most dramatically, oxygen levels fell to almost zero. A famous storm at Pitlochry in April 1974 brought with it rain with a pH of 2.4, which at the time was the most acid ever recorded in the UK. Rock pools in the area ended up with a pH similar to lemon juice; not all pH changes are produced by the creatures themselves.

Oxygen bubbles in a rock pool produced as green seaweed photosynthesises.

On windless days with heavy rain, a layer of freshwater may sit on top of the saltwater, causing some dramatic variations vertically within the pool. If you move your fingers through the water you will see the salt- and freshwater components swirling quite separately within the pool, though stratification of this nature is soon disrupted by wind.

In Chapter 2 we saw how conditions on rocky shores vary with the seasons. The same is true within a rock pool: for example, winter rain may flood it with freshwater and summer heatwaves can create pool conditions of such an extreme nature that even hardy shore creatures will be pushed beyond their tolerance limits and die.

The community of organisms found in rock pools will be affected by all the factors mentioned above, making each pool a unique little ecosystem. Further variation may be encouraged by chance events: erosion of the rocks altering pool shape and depth, rock falls, freak storms and associated sediment transport, the presence or absence of stones within the pool and so on. On the Ceredigion coast, in mid Wales, where we conduct a rock-pool survey every year, the shape and size of the pools, and their contents, differ tremendously every time we visit. This shore is home to a large population of Honeycomb Worms *Sabellaria alveolata*, remarkable animals that construct tubes from grains of sand and shell fragments to create large reefs from their adjoining homes. The size and shape of the reef varies from year to year and as a result impacts on the suite of pools that share that particular piece of intertidal real estate. Over the last four years the reef has expanded, to the detriment of our study pools, but since Honeycomb Worm reefs are rare and a UK Biodiversity Action Plan habitat, we don't mind. This species is found only sporadically from Lyme Bay on the south coast of England to the Scottish coast of the Solway Firth, and on parts of the Northern Ireland coast. As a species at the northern end of its range in the British Isles, it is sensitive to cold winters but may be on the increase as a consequence of climate change.

Unpredictable events and their effects on rock-pool contents make pool exploration all the more thrilling; you never know what you might find. To get a flavour of the differences between rock pools, try counting the number of species you can find in pools of a similar size and depth from each of the upper, middle and lower shore. It doesn't matter if you can't identify everything – just note the number of different types of organism. As we have seen, abiotic conditions in upper-shore pools fluctuate more than those in middle-shore pools,

Rock pools

An exposed-shore rock pool.

which in turn fluctuate more than those of the lower shore. As a result, the upper-shore pool should support fewer species than the middle-shore pool, with the lower-shore example being the richest.

In our experience, finding a pool or pools in each shore zone of similar size can be tricky, but there is a practical alternative. See if you can find a reasonably flat area of shore with a variety of different-sized pools, including really small pools and big pools if possible. Count the number of species in each. Here you might find that small pools, where conditions fluctuate more than in big pools, support fewer species; indeed, there might be a quite strong correlation between pool size and species richness for the whole range of sizes. However, bear in mind that we now have another factor affecting our result. Big pools might support a greater number of species simply because there is more room. Invariably there will be pools that buck the trend; here we are back to our previous observation that each pool has a unique set of conditions affecting it, and the community it contains will reflect this.

Pool specialists

Many organisms that live in pools may also live perfectly happily on the nearby shore, using pools as low-tide refuges, especially during unfavourable weather. But some life forms can be regarded as rock-pool specialists.

For grazing herbivores, rock pools are a little like 24-hour fast-food outlets; they are always open for business, and limpets, topshells and periwinkles can rehydrate at the same time as having a nutritious snack. These snails are partial to green seaweeds, which happen to do particularly well in pools as they are extremely tolerant of variations in environmental conditions. Depending on the grazing pressure, green algae such as the various *Ulva* species may thrive, often to the exclusion of other characteristic rock-pool seaweeds, including Serrated Wrack and Irish Moss. They do this by creating high pH levels and low concentrations of inorganic carbon (as a consequence of photosynthesis), which the other seaweeds cannot abide. However, limpet, topshell and periwinkle grazing reduces the abundance of green seaweeds and allows other, more unpalatable, species to flourish.

Two other pool specialists that benefit from herbivore activity are coral weeds *Corallina* spp. and pink paint weeds Corallinaceae. Both have high levels of calcium and magnesium carbonate in their cell walls, which effectively makes them inedible. As grazers remove other seaweeds, these unpalatable calcareous reds prosper. Coral weeds look like small, pink bushes with white tips and are usually found on the side walls of pools, whereas the pink paint weeds cover the sides and bottoms of many pools and look, and feel, like the rock itself. Their

A limpet and a Purple Topshell with various red seaweeds including pink paint weeds and coral weed.

colour varies tremendously; in well-lit locations the paint weed may bleach to white, but in deeper shady pools it may appear dark pink through to almost purple. Within the paint weed there are often oval patches of bare rock; these are the 'home scars' that limpets return to after feeding excursions.

Despite these red seaweeds being hard due to calcification, there are still some grazers, including chitons and the China Limpet *Patella ulyssiponensis*, that can cope with such tough food. All limpets feed using a tongue-like structure called a radula, which they scrape over the rock surface as they feed on various species of green seaweed, young stages of brown seaweeds and 'biofilm', which includes microscopic algae, bacteria, cyanobacteria, diatoms and protozoa covering the rock surface (see p. 98). The radula is essentially a ribbon covered in millimetre-long teeth that are made up of fibres consisting of an iron-based mineral called goethite, laced through a protein base. Because the mineral fibres are so thin, they are not prone to structural flaws that would weaken other materials, and their extreme strength is not dependent on scale. Normally, structures such as bridges and arches have to be of a certain size to achieve their desired tensile strength: not so limpet radula teeth. Their form is so strong that it can be compared to a single string of spaghetti holding up 1,500 one-kilogram bags of sugar. In fact, these teeth are made of the strongest biological material ever tested, even compared with spider silk (Barber *et al.* 2015). China Limpets exploit the strength of their radula teeth to eat pink paint weed, and in turn the paint weed has its growing point well below the surface – so, even with a growth rate of less than a millimetre a year, it can withstand limpet grazing.

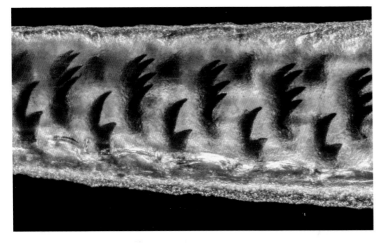

Limpet radula teeth and surrounding soft tissue.

Biofilm

Biofilm is a major food supply for grazing herbivores on the shore, but it is often impossible to see with the naked eye. It coats not only the rock surface but also the surfaces of seaweeds and invertebrate shells, carapaces and tubes. The composition of the biofilm will vary with height, season and exposure to wave action. For example, upper-shore films may contain a small amount of micro-algae but more in the way of cyanobacteria, and will occupy the rock surface alongside various tar lichens. Lower down, bacteria and micro-algae will be joined by microscopic seaweed holdfasts and eventually the sporelings of the seaweeds themselves. Biofilm composition will also vary with time; it is not long before bacterial and micro-algal films attract an array of single-celled animals and rotifers. These various biofilm inhabitants are also joined by the early settling stages of invertebrates, and the whole complex community may be bound together in a mucopolysaccharide film. These organic films are highly productive, and are understandably a very important part of the diet of intertidal herbivores.

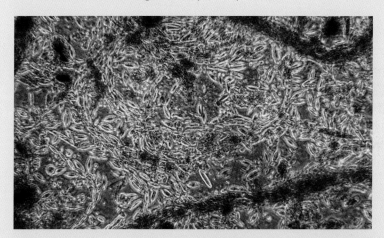

An extract from a biofilm sample that is approximately two weeks old. Magnification here is ×150, which allows us to see the abundance of benthic diatoms and bacteria (the smaller black dots).

Side view of biofilm, magnified ×5, showing the young seaweed sporelings beginning to grow.

A microscopic world

Coral weeds are also called 'fringe weeds' because they skirt the edges of pools. They are home to micro-communities barely visible except with a hand lens or a stereo microscope, which show that the strands of seaweed are covered in diatoms and organic matter. If you would like to investigate further, you might collect a little fringe weed, place it in a dish of seawater and gently tease it apart. Minute crustaceans swim out first, followed by an array of predatory worms and brittle-stars. Most amazing perhaps are the so-called sea-spiders which, though closely related to the arachnids, are not true spiders; some are just a millimetre or so long and move very slowly as they uncoil their long legs. They feed on the multitude of encrusting colonial creatures such as sea-mats, hydroids and tunicates that are also present on the coral weeds.

A few members of the fringe-weed community such as spiral tubeworms *Spirorbis* spp. are visible to the naked eye. They secrete a whorled tube of calcium carbonate on the surface of the wracks and coral weeds in which they live, with different species of *Spirorbis* being specific to their respective algal hosts. Within this microscopic world, these tiny worms feed by extending a fan of tentacles into the water, collecting plankton and particles of organic matter when immersed.

A fringe-weed community mainly consisting of coral weed and pink paint weeds.

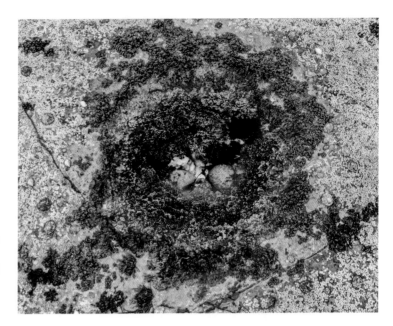

A rock pool showing distinct zones with coral weeds in the middle, then a green ring of various *Ulva* species, a brown ring of Pepper Dulse (actually a red seaweed) and then barnacles and limpets.

Perhaps the most unexpected and yet abundant rock-pool dweller is the larva of a midge, *Clunio marinus*. We do not tend to think that insects can live permanently immersed in seawater, and as a result marine biologists have largely ignored this interesting and diverse group. John Buxton, an entomologist, was the first person to describe this species almost a century ago, and nowadays it is recognised as one of 18 members of the genus around the world. *Clunio*'s larva lives in an irregular silk tube, dotted with debris, attached to the seaweed, and emerges to feed on encrusting diatoms and organic matter. Intriguingly this tiny creature has two biological clock mechanisms, enabling it to determine not only the state of the tide but also the phase of the moon, so that the male adult emerges on a receding spring tide. He needs to find a female pupa, release her from her case and connect to her with his genitalia. She is little more than a tube of eggs and will die shortly after mating and egg laying. The male will try to find other females with eggs to fertilise, but all this reproductive activity must be accomplished before the tide returns, at which time he dies. Without doubt these abundant chironomid midge larvae are important – and underestimated as a food source for other life in the pool.

While *Clunio* is often hidden within the fringe-weed community, rather more obvious in summer are the squirming hordes of tiny blue-grey marine springtails, *Anurida maritima*, which skate around on the meniscus of the rock-pool surface in a writhing mass. The

Larva of the marine midge *Clunio marinus* outside its tube, feeding on diatoms attached to some Green Branched Weed.

scientific name means 'wingless one that goes to sea' – very apt, as this animal is dispersed by the tide. An important shore scavenger and pool specialist, it aggregates on the water surface of sheltered pools because its entire body is covered in fine water-repelling (hydrophobic) hairs. As a flightless, air-breathing insect, less than 3mm long, it has a potential respiratory problem when the tide returns. A headlong dash to the safety of the splash zone every time the tide comes in would be like a twice-daily trek up the Himalayas. Additionally, staying on the water's meniscus at high tide could lead

Marine springtails *Anurida maritima* scavenging for detritus; note the hairs on the insects' bodies, which help them to repel water and use its meniscus for support.

to dispersal to the open sea or to land away from food and a mate. The solution for this diminutive insect is to find high-tide refuges in little air-filled crevices in the rock and subsist until emersion.

Anurida's hairy body facilitates the retention of a gas bubble around the animal during submersion. While it helps supply oxygen, the air bubble also acts as a compressible gas gill, like that found in water beetles in freshwater habitats. The gas bubble acts a bit like a lung, with gas diffusing into it from the surrounding water, thereby replenishing the air supply until the tide retreats. An inbuilt (endogenous) rhythm with a period of 12.4 hours allows *Anurida* to move in synchrony with the tidal cycle (McMeechan *et al.* 2000). This means that the springtail is tuned in with tidal movements and thus ready to seek suitable shelter before immersion rather than when it actually occurs. Forewarned is forearmed.

A truly remarkable, if minute, rock-pool specialist is a copepod called *Tigriopus brevicornis*. This tiny crustacean, a relative of shrimps and prawns, can only be found with close inspection, as it is less than a couple of millimetres long. It resides in upper-shore and even splash-zone rock pools in mid to late summer and can be recognised by its characteristic jerky swimming movements. Its orange-red colouration comes from a pigment called astaxanthin, a powerful antioxidant that protects the animal against ultraviolet light, which can be intense in such high-shore habitats. Astaxanthin can only be obtained from the animal's plant (phytoplankton) food, as it is unable to synthesise the compound itself. *Tigriopus* can withstand temperature fluctuations from −1 to +32°C, as well as a depletion of oxygen in the pool. Perhaps most remarkable is that its internal tissues can reach high salinities, up to three and a half times that of seawater. If the pool completely evaporates, *Tigriopus* burrows down into the drying salt and debris at the bottom and waits for the possible return of water (see Little *et al.* 2009 and Caramujo *et al.* 2012 for more details).

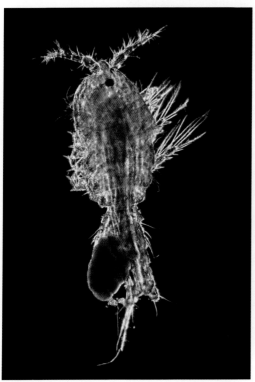

Tigriopus brevicornis, a copepod able to tolerate extreme salinity and temperature fluctuations in high-shore rock pools.

Diversity of life in a pool

On a calm day in summer, with the midday sun illuminating the water, many a budding rocky-shore explorer will have been enticed by the prospect of lying on the shore to gaze into a rock pool, only to find it disappointingly obscured by green seaweeds such as *Ulva* species. In winter, when most of our terrestrial environments shut down and lying on the shore might not seem quite such a tempting proposition, life in rock pools can be surprisingly abundant, as grazers from the surrounding area move in to the sanctuary of the pools to avoid freezing air temperatures. Most noticeably, the biodiversity of coral-weed and other fringe-weed communities reaches its annual peak at this time, and they can seem to be bursting with life.

Purple Topshells congregating in a rock pool for protection from cold winter air temperatures.

Ordnance in a rock pool is best reported and left to the experts to deal with.

As we will see below, many creatures are affected by cold winter conditions on the shore. A hard frost, accompanied by a low spring tide, can paint the upper shore white. Some species avoid these stresses by moving into deeper water for the winter, but plenty remain and tolerate the cold. Pools provide refuges because they are regularly warmed by incoming seawater. Take note, though, that all manner of organic debris will be brought in with the tide at this time of year, and searching these hidden worlds may require some care. Imagine the shock our students had when, as they moved wracks to one side in a large pool, a young seal pup reared up just centimetres away, hissing and showing its sharp teeth. Furthermore, it is surprisingly common to find ordnance, especially rockets, lying at the bottom of pools near military bases. There can be positives, however; some years ago on a sizzling hot day on the Dorset coast near Kimmeridge, we found a four-pack of lager in a deep, cool pool. From the condition of the dented cans they seemed to have been in the sea for some time – but, oh, their contents were refreshing.

Whichever season you choose to examine a pool, a slow approach is best. If possible, avoid casting a shadow, as many of the pool inhabitants will be scared into hiding. But don't worry if this happens: wait, and they will slowly reappear. For really deep pools, a crab line (available on any seaside promenade) works well, although all you really need is a tough thread with a small piece of bacon tied to the end. Lowering this into the pool and slowly retrieving it after a few minutes should reward you with crustaceans galore holding onto their meal. Drawing a small net slowly through the weed bordering the pool works well too.

Shrimps and prawns

The most familiar rock-pool species are the shrimps and prawns so beloved of small children equipped with hand nets. These animals can be extremely numerous – maybe 100 or so in a medium-sized pool – but tricky to spot as they are almost transparent and have pigment spots over their body surface that can change size to aid camouflage. So-called Chameleon-prawns *Hippolyte varians* have red, blue and yellow pigment cells, enabling the animals to change colour to suit the background: green against sea lettuce, for example, and red against red algae. At night, irrespective of habitat, the prawns change hue to a translucent blue, which is thought to protect them against predation. Hormones produced by endocrine glands located in the animal's eyestalks control this impressive process via the bloodstream.

During winter the atmospheric temperature may be such that upper-shore pools freeze, especially if subjected to freshwater runoff. Species such as the Common Prawn *Palaemon serratus* overwinter offshore, thereby avoiding cold rock-pool conditions until the spring, when they migrate to the shore.

Prawns are adept at walking, for which they use three of their five pairs of hind limbs; the fore limbs, equipped with fine pincers, forage for animal and plant debris and pass this material on to six pairs of feeding appendages. Like their close relatives the hermit and shore crabs, prawns are scavengers and are therefore important nutrient recyclers along the shoreline. They are also impressive swimmers, as anyone trying to catch a specimen by hand will know. When threatened, prawns flick their fanned tail segments under their body, causing them to dart backwards.

Prawns, like all crustaceans, have a tough, protective suit of outer body armour called a carapace that has to be shed and re-grown (see Chapter 9). Prawns complete the procedure (ecdysis) in about 30 seconds; mouthparts are hard enough to permit feeding after two hours and the process is complete within two days. Mating occurs while the female is soft, when she is said to be in her 'breeding dress'. The male, having completed his ecdysis earlier, places spermatophores under her abdomen, and soon afterwards the eggs are laid and fertilised as they pass over the spermatophores. The female cements up to 4,000 fertilised eggs to her swimming appendages in less than an hour, and then broods them for about four months until free-swimming larvae hatch and join the plankton. The planktonic larvae settle in late summer and are ready to breed the following year.

Sea anemones

Anemones are elegant additions to any rock pool, and two species in particular are often well represented: the Beadlet Anemone and the Snakelocks Anemone *Anemonia viridis*. Despite their common occurrence, there are many mysteries that remain to be solved about their biology. Although they look like little flowers, Beadlets are feisty, antagonistic individuals that fight each other in competition for space and food within the pool. The battle is a slow process for these animals without a brain. Each one rears up and head-butts the other in a contest that could last several days until the loser sidles off. Tentacles are used initially to size up an opponent, and if battle commences the Beadlet uses its 'beads' (technically known as acrorhagi) as the main weapons of offence. These beads (which give the anemone its common name) are packed with stinging cells, so as each combatant head-butts its adversary it is actually delivering quite severe stings. Pools have a finite space for these territorial anemones, and competitive disputes ensure that sufficient food is available for the victor.

Since individuals fight, it follows that they must recognise each other as different and worthy of aggression – but exactly how they

A Beadlet Anemone immersed in a rock pool with all its tentacles extended.

Detail of a Beadlet Anemone showing the animal's 'beads' or acrorhagi.

do this is a mystery. If Beadlet Anemones are produced as a result of sexual reproduction, then they should be genetically varied enough to be detected as 'other' when individuals come into contact. Interestingly, the extent and significance of sexual reproduction in Beadlet Anemones are at present uncertain. It used to be thought that sexes were separate, with both males and females brooding young, and offspring resulting from internal budding or parthenogenesis (reproduction from a gamete without fertilisation), or both. However, a recent paper suggests that at least some Beadlet Anemones are hermaphrodite (Costa *et al.* 2016). There is obviously a lot of work still to be done.

Common and widely distributed on all shores in northwest Europe, Beadlet Anemones come in many different colour morphs, from a wine-gum red through to quite pale greens and even shades of yellow. Recent DNA analysis (Douek *et al.* 2002, for example) indicates enormous variation within the population; what has been regarded up to now as one species may well be split into a number of different ones, and there may even be different species living in the different shore zones.

Beadlet Anemones typically eat small crabs, prawns and zooplankton, so the advantages of living in a pool are obvious. Any creature that brushes against the anemone's tentacles will be stung, paralysed and then eaten via the multipurpose orifice in the middle of the tentacle ring. Longevity of these animals is estimated to be in the order of three years in the wild, but Scottish advocate, antiquary and naturalist Sir John Graham Dalyell (1775–1851) kept a well-known specimen called

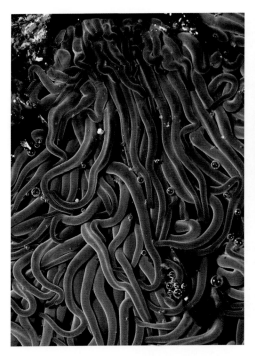 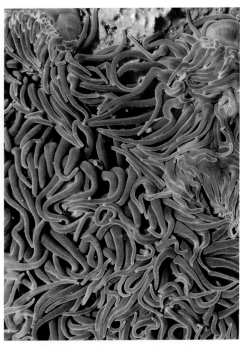

Snakelocks Anemone: green variants (left) with symbiotic algae in the tentacles (the bubbles are probably from the algal symbionts' photosynthesis); brown variants (right).

'Granny' in an aquarium for decades. Indeed, the anemone outlived him and spent its final years in the care of the Royal Botanic Gardens, Edinburgh, where it lived to be more than 50 years old.

Snakelocks Anemones also come in different colours, from a pale brown-grey through to an almost iridescent green with pink-purple ends to the tentacles. The green versions owe their colouration to chlorophyll-laden symbiotic algae (zooxanthellae) living in the animals' tentacles. Their geographic range is more limited than the Beadlet's; this is a western species extending along the northern coasts of Britain and along the Channel, but absent from most of the east.

Anemones with symbiotic algae have access to two food sources. The algae provide the products of photosynthesis, and Snakelocks Anemones consequently favour well-lit pools, but they also 'enjoy' a fish, prawn, plankton and even gastropod mollusc diet via the stinging cells in the long tentacles. The algal tenants benefit from their arrangement by having a place to live, and there is also a suggestion that the anemone provides nitrogen that the algae can utilise. Anemones with zooxanthellae often survive longer than their brown-grey cousins (Sabourault *et al.* 2009).

The Snakelocks Anemone, a pool specialist, usually has open tentacles, which create a large surface area and would cause the animal

Beadlet Anemones with tentacles retracted, accompanied by a Shanny and a Black-footed Limpet (the white tentacles visible around the shell margin and the straight edge to the shell confirm the identification).

to suffer from dehydration if it lived out of a pool. The Beadlet on the other hand is able to retract its tentacles fully within the body. The reduced surface area, combined with a mucous secretion over the body that helps to prevent it from drying out at low tide, means that this species can live just as easily outside pools; even so, the animal favours crevices and gullies rather than open rock. Some Snakelocks Anemones have been observed retracting their tentacles fully in aquaria, so it is physically possible for them to do so, but because their tentacles are so much longer it must be harder to accommodate them inside the column.

The dispersion of Beadlets and Snakelocks (the way individuals are arranged in space) is quite different, and this may well be a consequence of how the two species reproduce. Snakelocks are often found in dense aggregations so that they appear like a continuous mass rather than discrete individuals. They can reproduce sexually, but the preferred method is via longitudinal fission (splitting in two), which can take from five minutes to two hours. With fission, the next-door neighbours are effectively clones and will tolerate close proximity. The same tolerance does not extend to other species of anemone, and these are aggressively repelled.

If you can find a rock pool with aggregations of these anemones, try running your fingers gently across the tentacles of a Snakelocks hoard. The tentacles should react immediately, and you will feel a sticky sensation as the protrusions appear to suck on your digits. (There have been reports of some people reacting quite badly to Snakelocks stings, so it is best to err on the side of caution if you have sensitive skin.)

Snakelocks Anemones in the Channel Islands are often found with a colourful, symbiotic shrimp with blue stripes adorning its many appendages, called the Snakelocks Anemone Shrimp *Periclimenes sagittifer*. In 2007 the shrimp was discovered as a new record under Swanage Pier, Dorset, and then at Babbacombe, south Devon, in 2011. This apparent extension to the shrimp's range could be an indication that warming seas, as a result of climate change, enable it to live further north. However, its presence might simply be through discards from the aquarium trade, accidental or otherwise. It will be interesting to see if the species survives its recent range extension.

Since Snakelocks Anemones have sticky tentacles with a powerful sting, this raises the intriguing question as to why the Snakelocks Anemone Shrimp is not stung and caught by the anemone. Some insight has been gained by studies of the interaction between Snakelocks Anemones and crabs of the genus *Inachus*, to which it also plays host (Melzer & Meyer 2010). These crabs can be found living separately from anemones, and these individuals were stung when exposed to Snakelocks Anemones, but the anemone did not react to crabs that had been taken from another anemone. This suggests that prior exposure to the anemone results in some form of habituation, leading to the anemone no longer recognising the crab. The exact mechanism of this habituation is unclear, but it may involve transfer of mucus to the surface of the crab. If the crab's carapace is wiped with organic solvent, it is no longer protected from being stung (Weinbauer *et al.* 1982). Similar experiments have not been undertaken for the Snakelocks Anemone Shrimp, owing to the difficulty in obtaining anemone-naive individuals, and so the mystery is yet to be solved.

Other anemones that flourish in pools include the Gem, or Wartlet, Anemone *Aulactinia verrucosa*, which has small wart-like protuberances (verrucae) visible in rows (some white, others dark) when the anemone is closed up. This is a southern species reaching its northernmost limit

Rock pools

A trio of Gem Anemones in a rock pool.

Dahlia Anemones, showing some of the beautiful colour variants that exist.

in the Hebrides; its pink and white colouration affords it camouflage among the coral-weed fronds. The Dahlia Anemone *Urticina felina*, on the other hand, is widely distributed around Britain, where it thrives in areas exposed to wave action and penetrates to depths of 100m or so offshore. It can be common, but its presence in tide pools is nevertheless still a case for considerable excitement. Britain's largest intertidal anemone, it can be up to 120mm in diameter (a similar size to a grapefruit), and is spectacularly coloured, with pinks, reds and purples. Individuals typically have striped tentacles, but if these are retracted the anemone is still instantly recognisable by the shell and gravel fragments stuck to its column.

Fishes

As the tide recedes, many species of fish will use rock pools as bolt holes for protection. The Fifteen-spined Stickleback *Spinachia spinachia* survives in pools that reach low salinities in the upper shore, while the Butterfish *Pholis gunnellus* and Greater Pipefish *Syngnathus acus* are found in the middle and lower shore. At high tide these fishes are territorial and rarely travel far from their favourite pool.

Butterfish, aptly named for its slippery skin. The white-edged black spots are characteristic of this species.

The Shanny *Lipophrys pholis* is a real rock-pool specialist. Its body is soft and shiny because, as with many shore fishes, it has no scales, which would be damaged if they came into contact with rock surfaces. It is wise to be cautious if you wish to confirm this first-hand, as they do occasionally bite, have very sharp teeth and are often reluctant to let go; one of us has a sizeable scar on the palm of the hand to prove it. Mucus secreted by the skin is thought to help the fish slide over rough surfaces, and it may also help reduce desiccation while the fish is emersed. Why would a fish need to slide over rock surfaces when it could swim instead? One of the Shanny's strategies for remaining in position, when water movement may otherwise drag it out of a pool, is to hug the rock surface, keeping as much body contact with the rock as possible. The absence of a swim bladder stops the fish being buoyant. Shannies use their fins to brace the body against wave action, and the fin rays that regularly contact the rocks have thickened, protective cuticles. They typically dwell in crevices or holes in the rock, to shelter from the waves.

Young Shannies specialise in biting the legs off feeding barnacles, whereas older individuals favour small crabs, shrimps, prawns and various molluscs, and can be found with substantial amounts of seaweed in their gut. They forage over the shore area during high tide but return to the same pool (or pools) during emersion and may stay faithful to their chosen territory for many weeks. This behaviour is regarded as homing, and it must require a good memory and spatial recognition skills to enable the fish to recognise their pools' characteristic appearance and communities irrespective of tidal state.

In midsummer Shannies lay their eggs in pools, in crevices and under stones. They are guarded by the male, who fans them with his tail, presumably to keep a flow of water over them to aid with gas exchange and thermoregulation. A large female may lay up to 8,000 eggs in batches, and the male continues to keep an eye out for them as they hatch. If hatchlings stray too far he scoops them up in his mouth to spit them back into the nest of young. Crèche duty must be hungry work, however, and it is not unknown for larval fish to have appeared in male stomach contents. The young fish grow to about 20mm in the plankton and then settle out onto the shore, although they do not enter the breeding population for two or three years. Some individuals have been known to live for 14 years, although five years is a more common life span.

Brittle-stars, sea urchins and cushion-stars

Certain echinoderms also make use of rock pools; the Small Brittle-star *Amphipholis squamata* for instance can often be found in abundance amongst coral weeds. Crawling slowly over the fronds collecting dead animal and plant material on the seaweed surface for food, they are just visible to the naked eye. They also trap fine particles of detritus on mucus-covered spiny arms.

The Purple Sea Urchin has been recorded in southwest England, but it is on the edge of the Burren, on the west coast of Ireland, that the shallow, flat rock pools of the lower shore are covered from side to side in an abundance of this purple urchin. They use their teeth and spines to burrow into the rock, forming a depression in which they can hunker down, rarely needing to emerge. By holding seaweed and shell up high, like an umbrella, they can protect themselves from ultraviolet light. This 'masking', as it is called, along with living in a rocky hollow, also protects the body from loose material and crashing waves. These sea urchins graze on young algae, preventing seaweed establishment, leaving the habitat completely dominated by urchins.

Two cushion-stars are worthy of particular mention; both are limited to Britain's south and west coasts, where they are at the northern limit

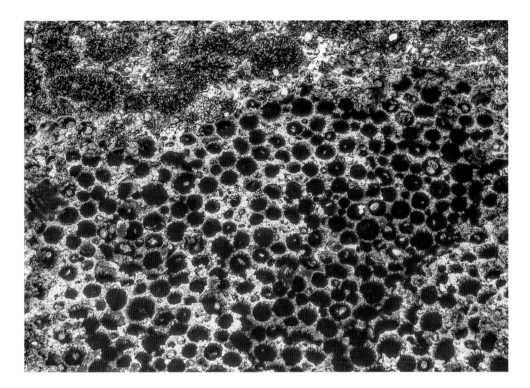

Purple Sea Urchins dominating a rock pool.

of their distribution. The first, the Common Cushion-star *Asterina gibbosa*, is relatively common and is found both in and out of rock pools. Those inhabiting pools tend to be olive-green, whereas those on the lower shore and in the sublittoral are orange or grey. Individuals can be up to 60mm in diameter, though 20mm is more common. These starfish change sex with age; they are male when they mature at about two years of age, and become female after four years. An individual can lay up to 1,000 little orange eggs in May in crevices or under stones, and after a few weeks these will hatch into tiny starfish. There is no planktonic phase; the young grow up close to the adults, and this proximity could lead to isolated (even genetically distinct) populations in separate pools. This cushion-star is known to scavenge on dead animal and seaweed remains, but its main source of food is the biofilm of organic material and bacteria that covers the rock surface. It feeds by everting its stomach (turning it inside out) through the mouth, pressing its lobes against the rock surface and then secreting enzymes to digest edible material before sucking everything back in again.

The Brooding Cushion-star *Asterina phylactica* is rarer and smaller, reaching a maximum diameter of only 15mm. It is impossible to distinguish this species from its commoner relative when young, but

Brooding Cushion-stars (top left and top right), with a clearly visible darker sub-star, and the Common Cushion-star, a larger animal without darker markings.

an adult will have a characteristic red-brown sub-star on a dark green background. It was first described in Britain by Emson and Crump in 1979 and seems to favour the sides and bottoms of stones in rock pools with good seaweed growth: its taxonomic type locality (where the species was formally recognised for the first time) is West Angle Bay, Pembrokeshire. Robin Crump, whom we met at the start of this chapter, became warden of the nearby Field Studies Council's Orielton Field Centre in the 1970s, and although retired for some years now he still makes regular visits to his cushion-star pool, meticulously counting and recording their breeding activity as well as population interaction with the Common Cushion-star. New populations have been discovered in suitable pools in St Bride's Bay and the Skomer Marine Conservation Zone, where the Brooding Cushion-star has also been recorded sublittorally. Other rock-pool populations have been recorded in Devon, Cornwall, Anglesey, Gower, Ireland and the west coast of Scotland.

In contrast to the Common Cushion-star, the Brooding Cushion-star is a simultaneous hermaphrodite when two years old, having briefly been a male for its first year. Adults congregate in groups of up to ten animals for egg laying, which probably favours cross-fertilisation, and, also in contrast to their close relative, they will stay as a group protecting the embryos until they hatch after about three weeks. Again there is no planktonic phase. This unusual behaviour explains the species' common name of Brooding Cushion-star. When the young disperse, the parents will often die.

The population of this rare cushion-star on West Angle Bay was reduced to a mere five individuals by oil pollution from the tanker *Sea Empress* in 1996. Fortunately, numbers have recovered to well over 1,000 in the intervening years. This recovery was no doubt helped by the cushion-stars' ability to self-fertilise, a useful strategy in an isolated habitat such as a rock pool.

* * *

If rock pools modify zonation patterns by acting as the oases of the intertidal, then it seems appropriate that we should next visit the true desert of the splash zone: here we enter the domain of the lichens.

A pesky crab latches on to a Victorian lady paddling through rock pools as the tide comes in.

Lichens: a primitive cooperative

chapter four

Light green; grey, brown and white; yellow; orange; black and greeny-black: lichens adorning the rocky shore provide enticing paint-like splashes of colour that can be observed while walking from the top of the splash zone through to the middle shore. The result is a series of distinctive bands that are distinguishable from a considerable distance.

Before embarking on that journey, however, let us begin by exploring lichens for what they are – fascinating organisms that are often overlooked in people's haste to get to the 'good stuff' lower down the shore.

Some background

Lichens have caused taxonomists severe headaches over the centuries; they were even classified as mosses for a time. Then in 1867 Simon Schwendener, a professor of botany and director of Basel's Botanic Gardens, announced to his scientific colleagues that lichens consisted of two organisms, a fungus and an alga. His 'dual hypothesis' was based on studies using nothing more powerful than a light microscope, and it took another 70 years or so before his ideas were accepted. Indeed at the time his hypothesis was treated with scorn by some of the most prominent lichenologists of the day (Honegger 2000); James Crombie, a Scottish reverend, jokingly referred to Schwendener's theory as 'a romance of lichenology, or the unnatural union between a captive algal damsel and tyrant fungal master'.

When we talk about a 'species' of lichen it is worth bearing in mind that this name refers to that of the fungal partner. Each

OPPOSITE PAGE:
Splash zone dominated by Sea Ivory (a fruticose lichen) and the foliose yellow lichen *Xanthoria* sp. Black tar lichen is visible in the distance above the grey barnacle zone near the water.

lichen consists of a unique fungus, and there are thought to be about 28,000 species worldwide, of which there are some 1,900 in the British Isles, including 450 coastal specialists. There are only approximately 100 algal species found in 'lichenised' fungi, each of which can be associated with a number of different fungal partners; they are also often found growing independently. The most common algae are species of the genus *Trebouxia*, which are found in over half of all lichens, including *Xanthoria* spp. on the shore. The orange-pigmented green alga *Trentepohlia* is found in about one-third of lichen species; it favours tropical and subtropical habitats as it is vulnerable to the cold but does occur in *Opegrapha* species including *O. rupestris*, a lichen that parasitises black tar lichens *Verrucaria* spp. (see pp. 127–128). Another related species, *O. physciaria*, occurs on *X. parietina*. In contrast, approximately 10 per cent of lichens have cyanobacteria partners instead; these are photosynthetic bacteria ('cyano-' refers to their blue colouration). Two common shore species, *Lichina pygmaea* (middle shore) and *L. confinis* (splash zone), have taken this evolutionary route.

The fungal partner never grows alone; when it is artificially induced to do so in a laboratory the resulting growth form is completely different from the normal 'lichenised' version and resembles nothing

Lichen zones on Skokholm, Pembrokeshire.

more closely than a shapeless mass. To complicate things further, lichenised fungi may contain different species of alga in different environments, or at different times in their life cycle, or even have both an algal partner and cyanobacteria. As the type of algal partner affects the growth form of the lichenised fungus, variants have been erroneously classified as separate species in the past. Nowadays such alternative forms are known as 'phases' of the lichen. Advantages for the fungal partner are obvious: it obtains sugars from algal photosynthesis. Frank Dobson, the author of the lichen bible *Lichens: an Illustrated Guide* (2011), suggests that the fungus actually stimulates the algal partner to produce extra nutriment. Advantages for the alga or cyanobacterium appear to be that they get a place to live where they are protected from environmental extremes and may have access to water and nutrients gathered by the fungus. Though this sounds perfectly plausible, there is still considerable debate as to whether it is true.

Whatever the exact details of this relationship, it must be a beneficial one, as lichens can live in some of the most extreme environments on earth, including cold polar habitats, baking-hot deserts, the tops of high mountains and below the high tide mark on rocky shores. Perhaps even more remarkably, lichens can live

Xanthoria ectanoides, showing leafy edges typical of a foliose lichen.

unprotected in the vacuum of space, with its widely fluctuating temperatures and cosmic radiation. In experiments led by Leopoldo Sancho from the University of Madrid, two lichen species, the Map Lichen *Rhizocarpon geographicum* and *Xanthoria elegans*, were exposed to such extraterrestrial abuse for 15 days (European Space Agency 2012). Attached to the outside of the Space Station, the lichens endured temperatures that ranged repeatedly from −12°C to +40°C. When the lichens were brought back to earth they were pronounced fully fit, with no discernible damage from their orbital adventure.

Growth in many lichen species is slow, less than a millimetre a year in some cases, but this is compensated for by extreme longevity; some species may survive for many hundreds of years. The apparently mutually beneficial, symbiotic, relationship is also an ancient one. The oldest lichen fossil in which both of the partners are clearly identifiable is 400 million years old, but there are suggestions of even older origins. Interpretation of possible remains is controversial, however, as lichens have no hard parts to facilitate fossilisation.

When two separate organisms are involved, reproduction and dispersal become rather complicated. Occasionally a piece of lichen will break off, and if it lands somewhere suitable it will continue to grow. Another non-sexual means of reproduction occurs through the dispersal of diaspores, specialised bits of the lichen that contain a few algal cells surrounded by fungus. How a 'bit of lichen that has broken off' differs from a 'diaspore' is apparently a bit of a grey area.

Only the fungal partner reproduces sexually, via the usual method of spore production. Most lichenised fungi belong to the Ascomycetes, and they produce spores in special structures called ascomata; some of these form the cups or plate-like discs on the top surface of lichens, which often aid significantly in the identification process. Once dispersed, the fungus must acquire the appropriate algal cells – and this is where the non-specificity of the algal partner is useful. One method of acquisition consists of the fungus parasitising another lichen containing suitable algal cells, and hijacking them for its own use. Even with the vast numbers of fungal spores produced, the chances of landing on a suitable substrate near an appropriate supply of algal cells are small. It is not surprising that vegetative reproduction is the dominant method used by lichens. The algal part of the partnership seems to have dispensed with sexual reproduction completely, and divides asexually within the fungus.

Lichen zones on the shore

On the rocky shore, lichens tend to occupy the bits that other species can't. The splash zone is a classic example: too dry for most marine species and yet too salt-sprayed for most terrestrial organisms. Here lichens occupy the zone almost exclusively, no doubt because they largely have no requirement for seawater. Immersion times increase as we follow our coloured bands down shore, and this, among other influences, affects the numbers and types of lichen encountered.

Light green

Starting right at the top of the splash zone, we find a beautiful light green lichen called Sea Ivory *Ramalina siliquosa*, which exhibits the fruticose growth form. Here the thallus, or body of the lichen, stands upright (or hangs down) from the base. This growth form exposes most of the lichen's surface area to the air, and fruticose lichens are consequently the most intolerant of airborne pollution. They also fare badly when subjected to water movement and wave action because their aerial parts are too easily damaged, which is why Sea Ivory is restricted to the upper parts of the splash zone. There are other coastal species of *Ramalina*, and, on rare occasions when southwesterly gales send salt-laden winds far inland, they can colonise inland sites. One rather famous example, some 50km inland from the south coast, is at Stonehenge.

Sea Ivory dominates the upper splash zone. Its fruticose growth form would be easily damaged by inundation with seawater.

Greys, browns and whites

Each year it is my (JA-T) privilege to help with an intertidal survey around Skomer Island National Nature Reserve, where we look at the full spectrum of rocky-shore inhabitants, including lichens. Our six survey sites vary from extreme exposure through to a very sheltered, shady, north-facing site called Hopgang. On the exposed, well-lit, south- and west-facing sites, with evocative names such as the Wick and the Pigstone, yellow and orange lichens abound, but on Hopgang the lichen flora is of a rather more subdued suite of colours. Shade-tolerant species thrive here; many are white or grey, but there is also a very attractive olive-brown foliose lichen called *Anaptychia runcinata*. 'Foliose' means that it has a horizontal, leaf-like thallus loosely attached to the rock and divided into lobes; most of the species in the sample area, by contrast, are 'crustose', which means, as the name suggests, that they form a flat crust on the surface. Common examples include Crab's-eye Lichen *Ochrolechia parella*, where the thallus and reproductive cups are the same creamy-white colour, and Black Shields *Tephromela atra*, where the grey lichen has black cups with pale rims. There are many others, including another *Opegrapha* species (*O. cesareensis*) with its *Trentopohlia* algal partner. Our Hopgang list includes 30 species within the sample area alone, more than I have ever seen on any other shore.

A muted colour palette of grey, white and brown lichens on a shady, north-facing shore, Hopgang, Pembrokeshire.

Lichens: a primitive cooperative

Anaptychia runcinata, a brown foliose lichen, contrasting with the yellow of Yellow Scales Lichen.

Black Shields with its distinctive white-rimmed black fruiting bodies (apothecia).

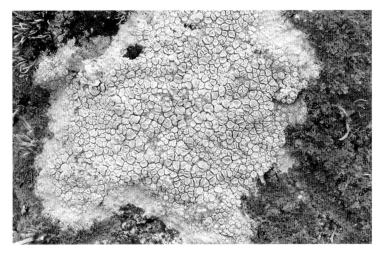

Crab's-eye Lichen, in which the fruiting bodies are the same creamy-white colour as the rest of the lichen.

Yellow

In sunnier locations, yellow and orange lichens come to the fore. In fact there can even be noticeable differences in the predominant colour of the lichen bands on two sides of the same bay. A shore in Pembrokeshire, Castlebeach, has a south-facing side and a north-facing one. The south-facing shore is noticeably more yellow and orange than the rather muted north-facing equivalent (at the same height above CD), which favours grey and black lichens.

Common yellow lichens and beautiful examples of the foliose growth form belong to the genus *Xanthoria*. In all the popular rocky shore guides the given species is the Yellow Scales Lichen *Xanthoria parietina*, which has obvious yellow cups peppering the thallus. In all probability another species, *X. aureola* (*ectaneoides*), is present as well, especially on Atlantic coasts. In the latter, reproductive cups are far less frequent. These *Xanthoria* lichens are relatively pollution-tolerant, favour sites enriched with bird droppings, and produce an orange-coloured pigment, parietin, in response to high UV-B levels, which is thought to give them protection from the sun's harmful rays. Parietin manufacture is enhanced by the products of photosynthesis manufactured by green algal cells (*Trebouxia* spp.) within the fungal body. Research directed at new products to protect human skin includes a number of substances derived from seashore lichens.

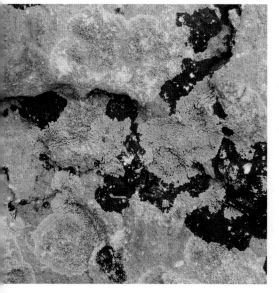

Two *Caloplaca* lichens: *C. thallincola*, with the long lobes and strong orange colour along the centre-line of the image, and *C. marina*, with a rather more diffuse structure, more muted in colour, around the periphery.

Orange

Below the yellow zone, with considerable overlap, is another group of 'sun-block' lichens, this time orange in colour. This bright band owes its existence to lichens of the genus *Caloplaca*. There are two common species, *C. marina* and *C. thallincola*, which are similar in colour, but *C. thallincola* has well-defined edges and lobes and prefers sheltered, shaded areas, whereas *C. marina* is more diffuse in appearance and favours sunny positions. Another slightly less common species, *C. verruculifera*, has a lemon-yellow tint in some specimens and is easy to spot once you get your eye in. These orange lichens can tolerate short periods of immersion in seawater, and their flattened growth form helps reduce damage from wave action.

Black

You might not think it likely that the world's media could be misled by a lichen, but during the *Sea Empress* oil spill in 1996 that's exactly what happened. Sent to cover the story, journalists from as far afield as Japan and Alaska stayed at Dale Fort. Having made long journeys from all points east and west, they arrived at the centre in the dark of a long February night. When they awoke the next morning they were horrified to see a universal black band at the top of every rocky shore in the vicinity. The immediate and hopelessly uninformed assumption was that this was oil; the breakfast room was awash with cries of disaster. As those who have read Chapter 2 will guess, the black band was in fact nothing other than a zone of the Black Tar Lichen *Verrucaria maura*. This crustose, matt-black lichen occupies the upper part of the upper shore and will intrude into the splash zone depending on local conditions, the vertical extent of its zone varying enormously with exposure to wave action.

As we will see in the next chapter, species classification is changing fast, a process ever invigorated by advances in molecular biology. These are exciting times for taxonomists. The 1981 second edition of Frank Dobson's lichen guide describes black maritime lichens called *Verrucaria maura*, *V. microspora*, *V. mucosa* and *V. striatula*. In the sixth edition (2011) there are colour pictures of *V. amphibia*, *V. ditmarsica*, *V. halizoa* (*V. microspora*), *V. maura*, *V. mucosa*, *V. degelii* (very rare and

Black Tar Lichen and *Caloplaca thallincola* close up. These flattened lichens are more resilient to water movement and wave action than their fruticose relatives.

Black Lichen accompanied by red encrusting algae, Green Tar Lichen, barnacles, rough periwinkles and a small shoot of Channel Wrack (bottom right).

confined to northern Scotland), *V. prominula* and *V. striatula*. Four species have become eight, and if there is a seventh edition in due course, more changes will appear. To prove the point, a recent literature search revealed a paper that discussed coastal lichens called *Hydropunctaria maura* and *Wahlenbergiella mucosa*, which we had never heard of. It seems the *Verrucaria* genus has been inspected and found wanting. To be clear, Black Tar Lichen *V. maura* is now known as *Hydropunctaria maura*. Green Tar Lichen, previously *V. mucosa*, is now *Wahlenbergiella mucosa* – and will be encountered when we reach the 'greeny-black' lichen zone.

Tar lichens are not the only black examples on the shore. A tufted, middle-shore lichen called Black Lichen *Lichina pygmaea* for instance is found on exposed shores in the barnacle zone; it does well on steep slopes in open sunny positions. Its cyanobacterium partner rapidly produces a chemical assisting in salt tolerance when immersed. Black Lichen also contains anti-grazing compounds, which probably accounts for the slightly medicinal smell it gives off if squeezed when damp. The TCP-like odour helps to differentiate this lichen from Creeping Chain Weed, a red seaweed it superficially resembles. The black tufts may harbour a variety of invertebrate guests, including an isopod, *Campecopea hirsuta*, that actually feeds on the lichen, the bivalve *Lasaea adansoni* (Chapter 7) and most excitingly a tiny little pseudoscorpion, *Neobisium maritimum* (Chapter 9), not to mention hordes of bright red snout mites Bdellidae.

Greeny-black

Green Tar Lichen occupies the lower half of the upper shore and often encroaches about halfway down onto the middle shore; in fact it is usually found lower down on the shore than any other lichen. On a sunny day it may appear 'greeny-black', but the colour varies and it can be difficult to tell apart from its up-shore neighbour Black Tar Lichen. The latter is invariably matt in appearance, whereas Green Tar Lichen often has a slight sheen and a thicker crust. In 2014, Niall Higgins (a botanist and plant scientist from the National University of Ireland, Galway) and some French colleagues produced a paper for the *Journal of Marine Ecology* looking at why Black Tar Lichen and Green Tar Lichen occupy different zones on the shore. They transplanted some Black Tar Lichen down shore and some Green Tar Lichen up shore. Both transplants died. It seems that Green Tar Lichen thrives in the moist (albeit salty) lower levels of the shore and is dehydrated beyond recovery higher up. In contrast, Black Tar Lichen tolerates dehydration better (the exact mechanism is still to be elucidated) but suffers from grazing pressure lower down the shore. Conventional wisdom would point to saltwater immersion times to explain seashore lichen distributions, but it seems there are other factors at play. Higgins *et al.* (2015) also looked at anti-grazer compounds (phenolics) in 13

Patches of Green Tar Lichen showing distinctively against the red of Old Red Sandstone, West Dale, Pembrokeshire.

different species of shore lichen and found a significant increase towards the lower shore; Green Tar Lichen had three times the amount found in Black Tar Lichen. It appears that grazing pressure might be a significant additional factor in explaining the distributions of rocky shore lichens, especially so for Green and Black Tar Lichens.

Other species

The main lichen zones on rocky shores are easy to spot from some considerable distance, but there are examples that require closer inspection, ideally with a hand lens, even to render them visible. *Collemopsidium foveolatum* (*Pyrenocollema halodytes*) is a case in point and probably the most common species in this genus; it appears as tiny little black-brown dots embedded in the shells of limpets and barnacles. Another species, *C. pelvetiae*, is found as brown dots on the thallus of Channel Wrack. Still more *Collemopsidium* species occur elsewhere on the shore, on different rock types, on shells and even on coastal dunes and calcareous sand; identification can be tricky.

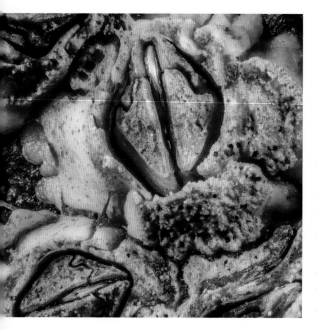

The ubiquitous black dots of *Collemopsidium foveolatum* can be found on barnacles and limpet shells on many rocky shores.

Lichen distributions – an overview

As we have travelled down the shore, through the different colour bands, we have seen lichen species composition change, but why? The simple answer is that competitively superior lichen species thrive in the splash zone where conditions are too harsh for other terrestrial organisms or marine ones to survive. Fruticose lichens such as Sea Ivory are typically able to over-grow Yellow Scales Lichen and other foliose species, which in turn are competitively superior to crustose forms such as Black Tar Lichen, and so on. Competitively inferior species had the evolutionary choice (so to speak) of either living in a progressively more extreme environment (for terrestrial lichens) as immersion times increased down shore, or becoming extinct. Other factors, such as light levels, herbivore grazing, moisture requirements, rock type and even the nature of the algal or cyanobacterial partners, also affect the outcome.

And then there were four

Throughout this chapter, we have presented the traditional view that lichens are a cooperative unit between a fungus and an algal and/or cyanobacterial partner (they can have both). Where two organisms share a long-term relationship with one another, this is referred to as symbiosis. This umbrella term covers mutualism, where both partners benefit, commensalism, where one partner benefits and for the other the relationship is broadly neutral, and parasitism, where one partner benefits to the detriment of the other. Lichens are typically thought to be paragons of the mutualism club, but it seems that there are complications: lichen examples have been found that exhibit parasitism, commensalism or mutualism, depending on the species involved or even the prevailing environmental conditions. Some cyanobacterial partners grow better in isolation than they do in a lichenised fungus.

Recent findings from the late 1990s onwards have discovered that other bacteria (apart from the cyanobacteria already mentioned), embedded in the lichen's structure, have functional roles to play in lichen biology. Suggested functions include nitrogen fixation, antifungal and antibacterial defence, and nutrient transfer from rock surface to fungal hyphae. Bacterial biofilms may also help with surface attachment in some lichens. So the modern view is that lichens are a symbiotic unit consisting of a fungus (the main structural component), an algal or cyanobacterial partner (or both) *and* a newcomer to the party in the form of additional bacteria. In other words, our symbiosis now contains three partners.

That might have been the end of the story, but a recent paper in the journal *Science* has added a new dimension of complexity and excitement to the debate: it appears there is yet another partner (Spribille *et al.* 2016). Many common lichens are also host to yeasts, which are embedded in their cortex. The implication is that the yeast partner may play an important role in the formation and control of a lichen's structure. Yeasts belong to the basidiomycete group of fungi, hence it seems that many lichens consist of *two* types of fungi, with an algal or cyanobacteria partner (or both), and other bacteria as well. Our symbiotic unit now contains four partners. These remarkable, ubiquitous, rocky-shore-enhancing organisms are little communities in their own right.

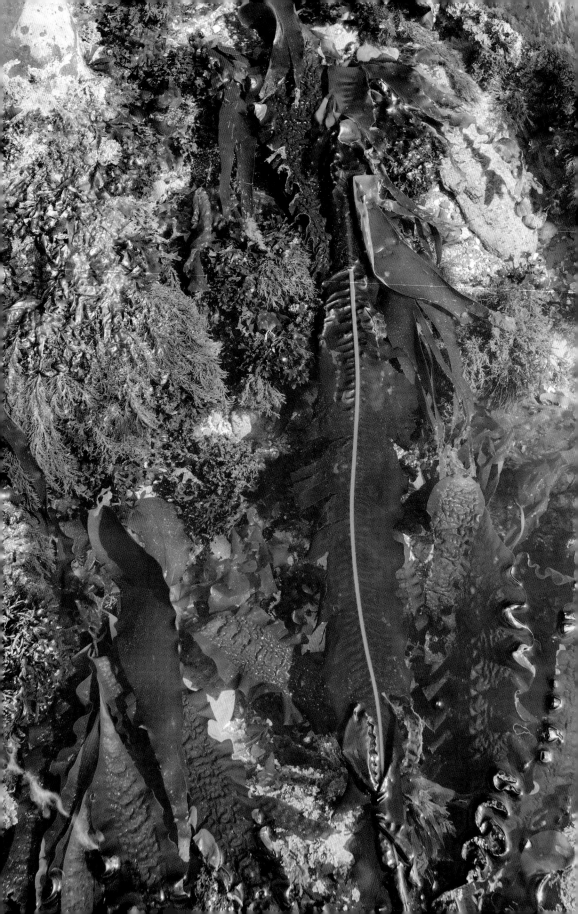

Seaweeds: the banquet that never was

chapter five

Defining seaweeds in the simplest terms, we might say they are marine plants, but up to about 450 million years ago *all* plants were marine, and all were algae. As plant life evolved to exploit the land it lost the ability to live in seawater, and there are now no marine species of mosses, liverworts, ferns or conifers. Of the 200,000 or so known species of flowering plants, only the sea grasses are truly marine; there are 50 species worldwide, and they account for less than 1 per cent of marine plants. Leaving aside the fungi and lichens (which are not really plants), almost all of the rest are algae.

Ninety-four per cent of marine plants are algae, but not all algae are seaweeds. In reality, most marine algae are single-celled and microscopic organisms; by 'seaweeds' we mean the macroscopic (visible to the naked eye), attached marine plants that form the most obvious 'plant' material on rocky seashores, especially those sheltered from wave action. If we start to look more closely at what a seaweed really is, the situation gets more complicated. Let's start by looking at human interaction with seaweeds, and how the study of them has developed over the past 300 years.

Algal exploitation

Human colonisation of the Americas, some 14,000 years ago, was from Asia by a land bridge connecting present-day Alaska to Siberia. The actual route south has been a subject of intense debate, but Tom Dillehay, from Vanderbilt University in the United States, along with colleagues from Chile, concluded that the migration route was along the

OPPOSITE PAGE:
A rock pool showing seaweed colours. The large browns are Sugar Kelp and Dabberlocks; bright greens are *Ulva*; the rest are various reds, including coral weeds, pink paint weeds, Dulse and Irish Moss.

Rocky Shores

RIGHT: Laver, one of the earliest types of seaweed to be eaten by humans.

BELOW: Dulse shows a distinct red colouration against the brown of the *Fucus* seaweed and bright green of the *Ulva*.

coast via a so-called 'kelp highway'. Remains of more than 20 different types of seaweed (*Porphyra*, *Gracilaria*, *Sargassum* and *Macrocystis*, for example) have been found associated with human settlements, implying a detailed knowledge and use of seaweeds as food and medicine. Traces of seaweed found on stone tools suggest this was a resource that was either ground or cut up for use (Dillehay *et al.* 2008).

Moving forward in time, we encounter seaweeds being used for food and medicine in China, Japan and India some 5,000 years ago. The legendary Chinese emperor Shennong (who is also credited with the discovery of tea) is said to have founded traditional Chinese herbal medicine, which involves exploiting the therapeutic effects of both land plants and seaweeds. The red seaweed laver *Porphyra* spp. was mentioned specifically by the Chinese as a foodstuff by AD 533, and there are references in early Greek and Roman accounts that include the use of seaweeds as fertiliser. Pliny the Elder (AD 23–79), author, naturalist, scientist, naval and army commander and personal friend of the Roman emperor Vespasian (an impressive resumé if ever there was one), documented 43 'remedies and observations' concerning the use of 'corals'. In those days corals were thought to be marine plants, and coral weeds *Corallina* spp. were included in the collection.

Closer to home, Scotland and Ireland have a long history of seaweed use for animal feed, fertiliser and human consumption; in the 12th century Irish monks gathered Dulse, a red seaweed, for distribution to the poor for emergency rations and food. In 17th-century Scotland, seaweeds, kelp in particular, were also used industrially to produce soda for the glass and soap industries. Following the decline of the soda industry, kelps were then used for the production of iodine until cheaper foreign sources became available. The British doctor Bernard Russell used brown algae to treat goitre and scrofula in the mid-18th century.

Seaweed science

In comparison to human exploitation of seaweeds, their scientific study is much more recent. A reasonable amount of published work and field study of seaweeds first appeared in the 16th and 17th centuries, and seaweed classification and identification progressed with the work of the Swedish botanist Carl Linnaeus (1707–1778), though he drew heavily on the work of those before him, such as Johan Jacob Dillenius (1684–1747), a German botanist who was Sherardian Professor of Botany at Oxford. Linnaeus included the algae in his Cryptogamia (plants with hidden reproductive organs) and named them using his new binomial system, which is the way scientific names of all organisms are presented to this day: Serrated Wrack, for example, is known as *Fucus serratus*, where *Fucus* denotes the seaweed's genus and *serratus* the species. Thus with this simple binomial arrangement we know something about the classification of the organism we have just named. All through the 18th century there was heated debate as to whether corals and coral weeds were animals or plants. Up to the mid-century the latter view prevailed, but by 1768 many authorities had come down on the side of animals. Less than a decade later the matter was finally decided in favour of 'vegetable material'. We now know of course that corals are animals and coral weeds are seaweeds.

One of the most distinguished algologists was William Henry Harvey (1811–1866), keeper of the herbarium and professor of botany at Trinity College Dublin. He travelled

Serrated Wrack, beautifully illustrated by William Henry Harvey.

A scrapbook page showing an assortment of seaweed from Jersey, c. 1850s. From the collections of the Natural History Museum, London.

extensively and wrote many definitive algal texts, including the three-volume *Phycologia Brittanica* and other equally impressive tomes for the Americas and Australia. He was the first authority to convince botanists to base seaweed groupings according to their pigmentation, although the original idea is credited to J. V. F. Lamouroux, a French marine biologist and naturalist, whose theory was ignored when first mooted in 1813. Harvey's groupings were Rhodospermae (reds), Melanospermae (browns) and Chlorospermae (greens).

The mid- to late 19th century saw the production of a number of seaweed books aimed at non-specialists. It is fair to say that the Victorians developed a bit of an obsession with seaweeds, spending hours collecting, drying and mounting them for decorative scrapbooks. In this way they demonstrated an increasing scientific knowledge of the natural world, adorning their homes at the same time. Indeed, at this time beachcombing was a fashionable pastime – especially, it seems, among ladies, who were often guided in their pursuit by the local doctor or curate. Margaret Gatty, a renowned children's author, was fascinated with marine biology and produced a popular guide to British seaweeds; her correspondence with William Henry Harvey must have helped in this endeavour. Amelia Griffiths of Torquay was such a prolific collector of seaweed material for Harvey that he dedicated one of his algal texts to her. He maintained she was worth a hundred of any other collector in the business. The seaweed genus *Griffithsia* was later named in her honour. Anna Atkins (1799–1871) produced what is considered to be the first ever book of photography; released as a series between 1843 and 1853, it was called *Photographs of British Algae*. Her specimens were placed on photosensitive paper and exposed to light, and a negative image of the seaweed resulted.

Margaret Gatty, children's author and beachcomber of repute.

In 1931, *A Handbook of the British Seaweeds* was produced by Lilly Newton (née Batten), professor of botany at the University College of Wales, Aberystwyth. The illustrations in this classic book are beautiful, and it was the first to offer a botanical key to aid seaweed identification, even though the species nomenclature is horribly out of date. The author included the algae in a now-obsolete grouping of 'lower plants', the Thallophyta, and then subdivided them into four groups using the names Cyanophyceae (cyanobacteria), Rhodophyceae (reds), Phaeophyceae (browns) and Chlorophyceae (greens). In 1962 the release of Eifion Jones's *Key to the Genera of British Seaweeds* was much needed as seaweed identification had moved on apace, leaving Newton's classic rather behind the times. Fittingly, Jones was a protégé of Newton's, and his contribution saw seaweed specialists through the hard times until the modern and definitive multi-volume series *Seaweeds of the British Isles* was produced by the British Phycological Society and the British Museum (Natural History) from 1977 onwards. Perhaps the closest modern equivalent to Newton's book is the recent Seasearch guide to *Seaweeds of Britain and Ireland* (Bunker *et al.* 2017), which uses excellent photographs and species descriptions to make seaweed identification as accessible as possible to the non-specialist.

Seaweed taxonomy continues to change rapidly, especially with advances in molecular identification techniques. Professor Mike Guiry, of the National University of Ireland, Galway, has said that a seaweed textbook is out of date the moment it is published, and his solution was to produce an online database that is constantly being updated: see www.algaebase.org for up-to-date names, publications, pictures of seaweeds and links to other useful resources.

Rocky-shore species identification past and present

Until recently, species identification relied exclusively on the morphological features of an organism such as size, shape, colour and detailed identification of parts using a hand lens or even a microscope. Identification guides or keys provided good-quality drawings, pictures or detailed descriptions to help. Often keys were (and indeed still are) structured so that the reader was led to the correct identification via a series of choices. In dichotomous keys there are two choices at every step, and each decision point in the key leads the reader to a new set of choices and eventually to a species name. Unfortunately, specimens that are young, old, damaged by grazers, undernourished, male, female, and so on can look very different from the norm. Moreover, sometimes correct identification relies on the specimen having seeds or other reproductive paraphernalia that may not be available. Considerable taxonomic experience may be necessary to identify a specimen correctly, even with the best available key to hand. Then there are the really perplexing situations where different species look identical and can only be told apart by dissection. There are also many cases where experts disagree about whether individuals belong to different species or whether they are just variants or ecotypes (the rough periwinkle group being a perfect example). And some, such as the Beadlet Anemone, may turn out to include a number of 'cryptic species' (distinct species that are erroneously classified under one species name).

Where morphological differences are impossible to detect, then other characteristics such as the species' life history, behaviour and physiology can be used to tell organisms apart. For example, it is accepted that in the rough periwinkle group *Littorina saxatilis* cannot be reliably told apart from *L. arcana* in the field. However, *L. saxatilis* females brood their young and often have young snails living around the shell opening before they disperse as 'crawlaways', whereas *L. arcana* lays pink, gelatinous egg masses in rock crevices from which the crawling juveniles emerge.

A modern solution to species identification is to use part of an organism's DNA from a standardised region of the genome (small sections of mitochondrial DNA in animals or chloroplast DNA in plants, for example). This technique originated in the 1960s, and there are currently two main approaches:

(1) DNA barcoding is a taxonomic method that uses a short sequence of an organism's DNA to identify it as belonging to a particular species. Once a DNA sequence is produced it can be used to identify an organism, like the barcode that is read by a scanner to identify products in a supermarket. Once a barcode has been obtained, it can be placed in a database for researchers to access, enabling them to identify unknown specimens.

(2) Molecular phylogeny uses sequences of DNA to determine evolutionary relationships between organisms. Every living organism contains DNA, RNA and proteins. Closely related organisms have similar molecular

structures in these substances. Mutations can occur over time, and if a constant rate of mutation is assumed then differences provide a measure of evolutionary dissimilarity or similarity. These data are used to provide a phylogenetic tree (a network of relationships) showing the possible evolution of various organisms. But how different do organisms need to be to declare them as separate species? Unfortunately this is open to interpretation. Different statistical approaches (of which there is a bewildering number) are used to make this decision; there are even computer programs to decide which statistical method is most appropriate. Differences and similarities are also dependent on the group of organisms in question. Human DNA differs from that of chimpanzees by 0.4 per cent, and that is considered to be different enough, in mammals, to conclude that they are separate species, but at present bacteria have to be 3 per cent different, or more, to be considered dissimilar, which probably under-estimates the number of species significantly. The figure is also currently 3 per cent in red seaweeds.

In short, there still seems to be a place for morphological, physiological and molecular genetic-based approaches to species identification. DNA analysis is just another (very useful) tool in the taxonomist's box.

Reds (Irish Moss), greens (sea lettuce) and browns (Oarweed): here, unusually, the colour differences are obvious.

What, then, are seaweeds?

So far, we have established that seaweeds are marine plants that belong to the larger taxonomic grouping called the algae. There are three main types of seaweed, namely reds, greens and browns. They share the intertidal and shallow sublittoral habitats, and as a result they have evolved similar structures. All photosynthesise using the light-harvesting pigments that give each group its distinctive colour, but does this mean they are closely related? In short, the answer is no. In one modern interpretation, reds, greens and browns are assigned to separate kingdoms of their own. In another, the reds and greens are part of the plant kingdom (Plantae), whereas the browns are separate, belonging to the kingdom Chromista, and are therefore not considered to be plants at all.

What, then, are the characteristics that are used to define whether or not seaweeds are related to one another? An obvious starting point is their pigments, which are housed in chloroplasts (the specialised structures in which the various photosynthetic processes occur). All seaweeds contain the chemical compound chlorophyll a, which is an essential component in the photosynthetic process. Green seaweeds add chlorophyll b as their main other pigment, and these two compounds give the greens their bright green colour, in common with land plants. Red seaweeds add phycocyanin and phycoerythrin as their accessory pigments, and the browns use fucoxanthin and chlorophyll c. The difference in pigmentation is far from the whole story, however. Greens, reds and browns produce different compounds to store the energy produced by photosynthesis. Each of the major seaweed groups stores materials in different locations, and their cell walls are made of distinct compounds. In addition, their chloroplast structures differ, as do some of the reproductive features: the green and brown seaweeds produce spores and gametes that move by means of hair-like flagella, but these flagella are lacking in the reds. Perhaps the most fundamental differences between the three seaweed groups concern their subcellular structure, and this leads us into some of the most controversial territory in modern biology.

In 1910, Russian botanist Konstantin Mereschkowski came up with a revolutionary idea concerning the evolution of the eukaryotes. These are organisms, including fungi, plants, seaweeds and all animals, whose cells contain a nucleus and other organelles (specialised subunits within the cell including mitochondria and chloroplasts) sheathed by a membrane or membranes. The other major group of organisms is the prokaryotes, which contains the earth's most ancient life forms. In this latter group of organisms (all unicellular bacteria and cyanobacteria), the cell contents have no membrane-sheathed organelles and no nucleus. Konstantin's remarkable idea was that the organelles within eukaryote cells used to be free-living prokaryotic organisms that were engulfed by the 'host' and eventually became part of the host's cellular apparatus; life continued, to the mutual benefit of both partners. The term for a long-term relationship between different species such as this is symbiosis (from the Greek *syn*, together; *biosis*, living), and we encountered classic examples of this evolutionary strategy in the lichens (Chapter 4). In the particular case of one partner living within the other, having been engulfed by it, the relationship is termed endosymbiosis.

At the time, in the early 20th century, there was no way of validating Konstantin's idea, as the genetic know-how and technology did not exist. In any case his idea was considered preposterous and promptly discounted by his peers; his opposition to Darwinian natural selection probably did not help his cause. Jump forward some 50 years to research conducted by Lynn Margulis (1938–2011), a remarkable evolutionary theorist working at Boston University. She championed Mereschkowski's idea, in turn receiving ridicule and indeed vitriol from scientific journals that refused to publish her work. Like Mereschkowski, she had a different take on 'evolution by natural selection' as well. Because of her work on symbiosis she opposed competition-oriented views of evolution, so beloved of neo-Darwinists such as Richard Dawkins, and stressed instead the importance of cooperative relationships between organisms in the evolutionary process. Her later work included the co-development of James Lovelock's famous Gaia hypothesis that the earth functions as a single self-regulating system (organism). Margulis and her son and co-worker Dorion Sagan argued that 'life did not take over the globe by combat but by networking'. Public and rather robust disagreements with neo-Darwinists continued, but in 1978 the fact that mitochondria were descended from free-living bacteria and that chloroplasts were descended from cyanobacteria was demonstrated for the first time. In the early 1980s studies of seaweed chloroplast genes showed that the chloroplast DNA was cyanobacterial DNA and bore little resemblance to that of the host.

So, to return to our original quest of sorting out what seaweeds are, we are now in a position to understand why red and green seaweeds are considered to be significantly different to the browns. Red and green seaweeds (and land plants) have chloroplasts that are surrounded by two membranes, suggesting they developed directly from endosymbiotic cyanobacteria. In other words, a single-celled ancestor engulfed a free-living cyanobacterium that eventually became the host's chloroplast. This significant event in evolutionary biology is thought to have occurred about 1.2 billion years ago. In browns, the chloroplasts are surrounded by *four* membranes. It seems, therefore, that the chloroplasts in brown algae were acquired when the host cell engulfed another free-living organism that already contained a chloroplast (had already undergone endosymbiosis). One school of thought maintains that the free-living organism that was engulfed was a red seaweed. Because this represents an endosymbiotic organism within another, this becomes secondary endosymbiosis. This event happened rather more recently; estimates vary, but 150–200 million years ago is the generally accepted timeframe.

Why are seaweeds different colours?

In Chapter 2 we saw how different wavelengths of light penetrate to different depths in coastal waters such that red light is absorbed first, blue light next, leaving green light penetrating to the greatest depth. A theory to explain how seaweeds might respond to these changing wavelengths was first proposed in 1883 by Professor T. W. Engelmann (1843–1909), a German botanist, physiologist, microbiologist and musician. According to this theory, green seaweeds, which absorb mostly red light, live in the shallows, brown algae, which absorb mostly blue light, occupy the next zone down, and finally red seaweeds, which absorb predominantly green light, penetrate to the greatest depth. This elegant proposition was even given a rather stylish epithet, that of 'chromatic adaptation'. There was a problem with this approach, however, which most people ignored at the time – and that was that the *quantity* of light also changes (decreases) with water depth. It would be difficult to know whether it was the quality (wavelength) of light that caused the observed changes in seaweed distribution or its quantity. Friedrich Oltmanns, another remarkable pioneer in the seaweed ecology world, said as much in 1892, but his notion did not engender much support. However, some conclusive evidence can be gained by observations in submarine caves in which the depth of the water is constant (and hence there is no change in the wavelength of light from front to back): at the mouth of such caves green seaweeds flourish, further into the cave brown seaweeds predominate, and at the back there are red seaweeds exclusively. It appears that the wavelength of light may not be as important as the quantity. Red seaweeds are shade-tolerant. So Oltmanns' objection has been sustained – but you wouldn't think so if you looked at many modern-day texts: they still peddle the erroneous chromatic adaptation idea.

Brown seaweeds

Brown seaweeds are, generally speaking, the big seaweeds on the shore. There are estimated to be around 1,800 species worldwide, about 200 of which have been recorded from the British Isles. As a group they favour colder, nutrient-rich seawater, so they are ideally suited to the North Atlantic and North Pacific. Very sheltered shores may be covered with a thick blanket of weed such that the underlying rock is invisible, and this canopy can provide shelter for other species of seaweed, especially the shade-tolerant reds. Many other species of seaweed and animals may live on the surface of brown seaweeds, and the algae thus provide important extra habitat for a variety of epiphytic forms. The largest brown seaweed in the world is the Giant Kelp *Macrocystis pyrifera*, a Pacific species that may grow to 70m in length and achieve 50–60cm of growth per day. British brown seaweeds are rather modest in comparison, but we have seen kelps and Egg Wrack a couple of metres long in our travels around Britain, and a colleague and friend of ours, Dr Stephen Morrell, has seen 6m lengths of Wireweed.

Egg Wrack en masse.

The various brown seaweed species have characteristic zonation positions on the shore. Channel Wrack can live further up than any of the other common brown seaweeds, and many hypotheses have been presented to explain how it can survive this high. Another fundamental question is why it is there in the first place. As to how, it was assumed that upper-shore species of brown seaweed lost water more slowly than their lower-shore counterparts. Early experiments, however, ignored surface area to volume considerations, and when these were factored in it was concluded that all brown seaweeds lose water at approximately the same rate. It appears, in fact, that Channel Wrack does not avoid dehydration but tolerates it, and can recover from 96 per cent water loss with minimal damage, which is truly remarkable (Schonbeck & Norton 1979). An excellent account of the factors controlling seaweed zonation, which refers to more of Schonbeck and Norton's work, can be found in Moore & Seed's book on *The Ecology of Rocky Coasts* (1985).

As its name implies, the thallus (body) of Channel Wrack is channelled, and another suggestion was that water was retained in the channel when the weed was emersed. However, this failed to take into account that any advantage gained from having a concave surface pressed to the rocks (the natural position the seaweed adopts when out of water) was more than offset by having the convex surface open to the air. The concave part of the thallus most likely provides a humid microclimate where the developing zygotes can lodge during the vulnerable early stages of growth, before they fall out and attach to the rocks.

Recently it was discovered that Channel Wrack hosts a fungus, *Mycophycias ascophylli*, within its tissues, and researchers claim that this symbiont is largely responsible for the seaweed's remarkable ability to tolerate environmental extremes (though the exact mechanisms are still being elucidated). Channel Wrack has become so specialised, spending up to 90 per cent of its life out of seawater, that if kept permanently submerged it dies. It may well be that the fungal part of the partnership cannot tolerate this level of immersion. We saw in Chapter 4 that lichens are a symbiotic unit where the fungus provides the structure that algal cells live within. Here in Channel Wrack we seem to have a 'reverse-lichen', where the alga provides the home that the fungus lives in.

Brown seaweeds have been part of the biosphere for millions of years so are well adapted to the average conditions they might

Channel Wrack, typically the highest brown seaweed on British shores.

encounter on the shore. It is the extreme climatic events that seem to cause most harm, trimming back the edges of an organism's vertical range and reinforcing zonation patterns in the process. Seaweeds may suffer lethal damage on exceptionally hot summer days, especially during neap tides, when seaweeds on the upper shore may be left uncovered for days at a stretch. It is not only the temperature that kills, but also the length of time a species is exposed. In addition, damage seems to be worst on warm humid days rather than on hot dry ones; rapid drying of the tissues appears to protect the weeds against lethal temperature effects.

Another real issue with high-shore living is food supply. Seaweeds only undergo gaseous exchange and absorb nutrients when immersed; photosynthesis shuts down rapidly on emersion. In the brief periods that Channel Wrack spends under water it must recover rapidly from any dehydration effects and start photosynthesising before it is high and dry again. Experiments have demonstrated that this remarkable seaweed can be photosynthesising at its full rate within 20 minutes of previously suffering severe desiccation (Dring 1986).

If the common brown seaweeds of the British Isles are cultured under water in the laboratory their growth rates correlate exactly with their zonation position (Moore & Seed 1985). Channel Wrack grows more slowly than Spiraled Wrack, which grows more slowly than Bladder Wrack, and so on. The situation in real life is more pronounced, as high-shore weeds spend less time under water, when photosynthesis can take place. Not only is growth rate slower in absolute terms, but also there is less time available for growth.

Even casual inspection of a sheltered shore will reveal that brown seaweeds get longer as you move towards Chart Datum. In experiments where Spiraled Wrack was removed from the rocks, the lower limit of Channel Wrack subsequently extended down the shore (Moore & Seed 1985). If Spiraled Wrack was then allowed to grow, it soon out-competed Channel Wrack, and the normal zonation pattern was re-established. It seems that each species of brown seaweed could thrive lower down the shore than its normal position would suggest but is prevented from doing so by a faster-growing neighbour living just below it. But a more careful look at brown seaweed length will reveal another interesting fact. Even within, say, Channel Wrack's vertical range, individuals get longer as you move down the zone. In other words, the increase in length is not just because of the change in species, it happens within a species as well, and for the same reasons.

Seaweeds' ability to tolerate extreme environmental conditions is not constant; it changes as the season progresses. Seaweeds become hardened to tolerate dehydration much as land plants become more frost-hardy with time. Upper-zone individuals are hardier than their lower-zone relatives.

Moving down the shore from Channel Wrack's domain, the next seaweed we encounter is Spiraled Wrack. Similar figures for dehydration tolerance as Channel Wrack are quoted in the literature, but direct observation tells a different story: Spiraled Wrack is almost always found just below its high-shore neighbour. It could be that this position relates to the species having a greater surface area from which to lose water, but differences in nutrient uptake may also play a part. Since Spiraled Wrack can outgrow Channel Wrack under constant conditions, it follows that it must be a more competitive species that is good at seizing nutrients when they are available. Perhaps this ability means it is less able to tolerate the more nutrient-poor environment of the high shore. This state of affairs repeats itself with other species of brown seaweed as you move down the shore. Overall, the coastal story is similar to the situation in terrestrial

Seaweeds: the banquet that never was

Spiraled Wrack usually occupies the zone just below Channel Wrack (top left).

meadows, where many species can coexist when nutrients are in short supply but competitive coarse grass species take over following the addition of fertilisers.

Typically one might encounter Bladder Wrack next in the sequence down the shore after Spiraled Wrack. This species can only recover from a maximum of 40 per cent water loss but, like most seaweed species, it manages to offset dehydration sensitivity when the tide retreats by collapsing into a heap (seaweeds do not possess the internal supportive structures of higher plants); a seaweed in a heap will only lose 20 per cent of the water it would otherwise lose if laid out flat. Unusually among seaweeds, Bladder Wrack is also able to tolerate living on exposed shores, as it can change its growth form when confronted with increased wave action: the exposed-shore form *Fucus vesiculosus* var. *linearis* lacks the gas bladders that would create drag in rough seas.

Bladder Wrack, a characteristic middle-shore seaweed.

Generally occupying a similar zonation position to Bladder Wrack is the long-lived brown seaweed Egg Wrack. Most fucoids live for between three and five years, and many types of kelp survive rather longer, but Egg Wrack can survive for more than 25 years. Per Åberg, professor of marine ecology at the University of Gothenburg, suggests that it could survive by re-growing from the base for more than a century (Åberg 1992). If you ever find yourself on a shore where there is an abundance of Egg Wrack, tread carefully – you may be surrounded by very old seaweeds indeed. The age of an Egg Wrack individual may be estimated by counting the seaweed's bladders. You can assume as a rough guide that they grow a bladder a year, but be mindful that bladders and fronds may be lost if the individual has been damaged. Egg Wrack may be slower-growing than some of its mid-shore contemporaries, but it is shade-tolerant and persistent; when surrounding shorter-lived species die off, Egg Wrack can take over and then prevent re-colonisation under its dense canopy.

Egg Wrack is often host to a red seaweed, Wrack Siphon Weed; the red penetrates the host's tissues, implying the relationship is parasitic and that the red gets sugars from the Egg Wrack. Wrack Siphon Weed still has its photosynthetic pigments, though, so it is classed as a hemi-parasite rather than a fully dependent one. There are suggestions that sugars are transported both ways in this relationship, however. Wrack Siphon Weed in turn is generally infected with a tiny, more or less colourless parasitic red alga called *Choreocolax polysiphoniae*, to which it is closely related.

Egg Wrack, with Wrack Siphon Weed attached to its surface.

It is tempting to see hosts such as Egg Wrack as hapless victims providing inviting surfaces for all sorts of epiphytic and parasitic organisms to make use of, but in fact this brown seaweed has a toxic surface that exudes antifouling chemicals, polyphenolics, in its defence. Wrack Siphon Weed often attaches to damaged areas of the host where the defences are weaker. Intriguingly, recent work has revealed that seaweeds are also active combatants in the face of invertebrate attack. Egg Wrack exudes anti-grazing chemicals such as tannins that deter flat periwinkles, one of the few snails that eat adult browns; the presence of these snails in the water is enough to stimulate Egg Wrack to increase its tannin production (Flöthe & Molis 2013). Damaged-tissue exudate also stimulates neighbouring seaweeds to increase their production of anti-grazing chemicals. Amazingly, seaweeds are communicating with each other.

The next stop in our journey down the shore takes us to Serrated Wrack, a seaweed that is common on the lower shores in both sheltered and exposed locations. It is similar to Bladder Wrack in that a 40 per cent threshold for dehydration recovery has been quoted, but its growth rate far outstrips that of its bladdery companion. The serrations that give this seaweed its name cause micro-turbulence in the water surrounding the fronds, which helps oxygenate the water. Individuals growing on sheltered shores have more serrations, which helps offset the dangers of more stagnant conditions; on exposed shores a reduction in serrations helps reduce the seaweed's surface area, making it less vulnerable to drag.

Serrated Wrack, a lower-shore brown seaweed.

Rocky Shores

ABOVE: Thong Weed: the strap-like thongs are the reproductive parts that emerge from the centre of the button-like growths shown towards the bottom of the image.

RIGHT: Sugar Kelp is abundant on this lower, sheltered shore (BES grade 7).

Another low-shore brown worthy of mention is Thong Weed. Young forms of this species are like brown buttons, up to 20mm across, with a concave upper surface. A branched, strap-like frond grows out from the button after about a year; this is the reproductive phase, and it dies off after gametes are shed, usually in late spring or summer. The lower surface of the buttons is often home to a number of epiphytes, spiral tubeworms for example, but the concave upper surface remains free from squatters.

If you are lucky (or organised) enough to visit a rocky shore when there is a low spring tide, you can experience the magnificent kelp forest emerging from the depths. These brown algae are the trees at the base of the intertidal; they grow more quickly than their up-shore relatives, attain a greater stature and cast a dense shade, which means that the 'woodland floor' is host to shade-tolerant species only. Sugar Kelp is characteristic of sheltered shores, Oarweed and Dabberlocks of more exposed areas. Desiccation tolerance in these intertidal giants

LEFT: Oarweed, with strong, flexible stipes and fingered blades that help it cope with wave action on exposed rocky shores.

BELOW: Oarweed holdfasts, a potential microhabitat for epiphytic seaweeds and various invertebrates.

is down to 20 per cent, and these seaweeds have a much larger surface area from which to lose water.

Algae, like land plants, grow from active areas of cell division called meristems. The Wracks mentioned previously grow from the tips, meaning that the youngest part of an individual is at the end of the frond. In the kelps, however, the meristem is located between the blade and the stipe (stem), so harvesting need not kill the seaweed, as long as it is done carefully and only the blade is taken. With this growth system, the oldest part of the blade is at the end. Sugar Kelp can grow up to 20mm per day, and the length of the blade is the result of the interaction between new growth at the base and damage to the tip. Length increases between January and July as a result of the extra light stimulating growth and calm seas reducing damage. Between August and January, growth slows and damage increases, especially during autumn gales, meaning that the blade is at its shortest in winter. If you ever find kelp plants washed up in the strandline, it is worth having a look at the cross section of the stipe; you might be able to age an individual from the number of light and dark rings. The former denote rapid summer growth, the latter the slow growth during less productive times of the year.

Seaweeds are held to the rock mechanically by a holdfast (hapteron) with minute outgrowths that exploit cracks and crevices in the rock structure, and chemically with the aid of a range of adhesive compounds: polysaccharides and proteins, for example. By using both mechanical and chemical means, some quite sizeable plants can be held in place by diminutive holdfasts. In the wracks the holdfast approximates to a disc shape, but in kelps the structure is branched, and can in itself provide valuable habitat for a range of rocky-shore invertebrates and indeed other seaweeds (see photograph on p. 151). A kelp holdfast might house sponges, sea-squirts, bryozoans, hydroids, amphipods, snails, crabs and a host of epiphytic seaweeds. One study clocked up an impressive 99 different types of organism on one Forest Kelp *Laminaria hypoborea* holdfast (Little *et al.* 2009).

The sex lives of seaweeds are surprisingly complicated and involve a bewildering array of specific terms. As is often the case, complex terminology masks an extremely interesting underlying biology. A brief look at seaweed reproduction will suffice here; more detailed treatments abound in the literature and online, but they are not for the faint-hearted. Brown seaweeds have two phases to their life history (called 'alternation of generations', which also occurs in terrestrial plants), and the two phases often look completely different. In the past, different phases of the same seaweed's life cycle were erroneously named as separate species. In kelps the life history is such that the typical individual seen on the shore is something called a diploid sporophyte (diploid: chromosomes in pairs in the nucleus; sporophyte: spore-producing stage), while the other phase, a haploid gametophyte (haploid: single set of unpaired chromosomes in the nucleus; gametophyte: gamete-producing stage) is microscopic (see p. 153). In Chipolata Weed *Scytosiphon lomentaria* (it really does look like a string of sausages), the recognisable individual is the gametophyte, whereas the sporophyte is a brown crust on the rock. In the common fucoid seaweeds, Bladder Wrack for example, the sporophyte phase is the characteristic individual, while the gametophyte takes the form of special reproductive cavities (called conceptacles) within the tissue of the sporophyte (see p. 154).

Seaweeds: the banquet that never was

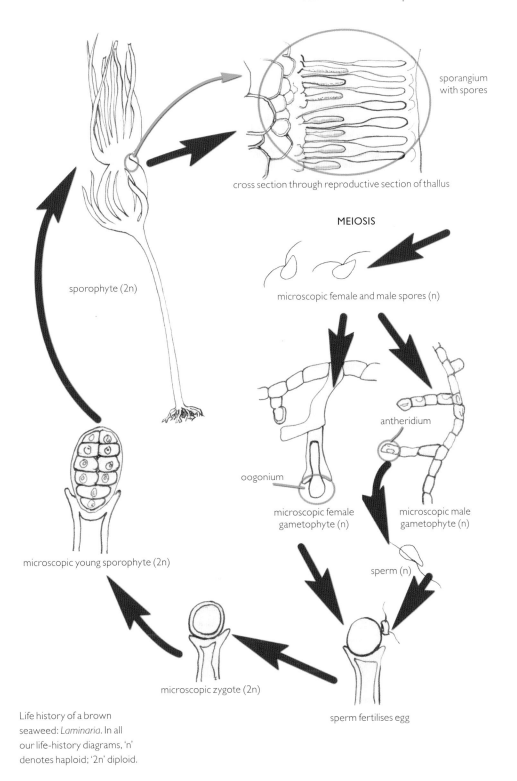

Life history of a brown seaweed: *Laminaria*. In all our life-history diagrams, 'n' denotes haploid; '2n' diploid.

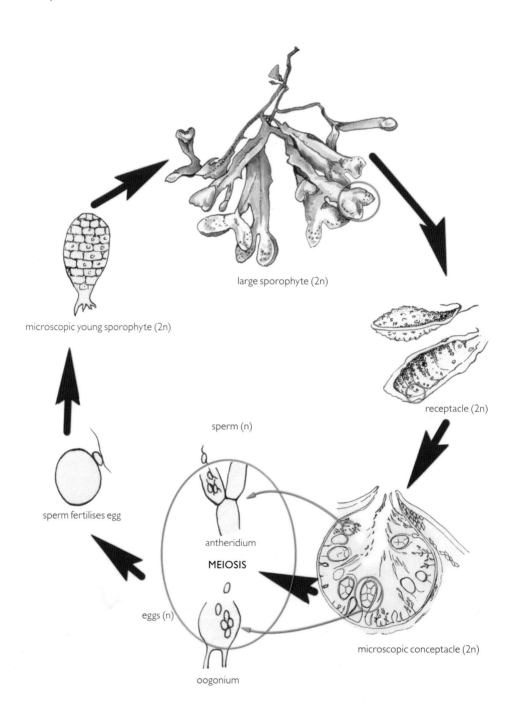

Life history of a brown seaweed: *Fucus*.

Red seaweeds

Red seaweeds are the oldest known seaweed group. They are closely related to the green seaweeds and are still regarded as plants, at least in most classification systems. Fossilised remains of one particular species, *Bangiomorpha pubescens*, found in rocks in Canada, have been dated to a mind-boggling 1.2 billion years ago. On a recent shore survey closer to home in West Angle Bay, Pembrokeshire, we were excited to find a minute piece of bright red fluff sticking out of a rocky crevice; subsequent sectioning and microscopic identification confirmed that this was a modern-day relative of *B. pubescens* called *B. atropurpurea*, which strongly resembles its fossilised ancestor. It was genuinely awe-inspiring to be in the presence of a life form with such an ancient lineage, particularly since seaweed fossils are so rare (in most cases they lack any hard parts that lend themselves to fossilisation). There may be other ancestors that were just never fossilised, or perhaps that exist out there somewhere still awaiting discovery.

Today, there are about 7,000 species of red seaweed worldwide, and some 350 of these have been recorded around Britain, Ireland and the Channel Islands. The pigments enable photosynthesis at low light levels, such as at the base of the kelp forest; the downside to this arrangement is that they are damaged if exposed to strong sunlight. The upper shore is therefore not a place where one would expect red seaweeds to thrive, but there a few species that may be encountered.

A species typical of this zone is Tough Laver *Porphyra umbilicalis*: the common name is apposite. This remarkable seaweed has been the subject of recent research into the biochemistry of stress tolerance in intertidal seaweeds. In a 2008 paper, Priya Sampath-Wiley and colleagues from the University of New Hampshire describe how Tough Laver overcomes light damage, heat stress and desiccation by increased antioxidant production. Tough Laver is also able to alter pigment levels seasonally so that in the winter, when light levels are low, pigment (e.g. phycobilin) production is increased to capture enough light to drive electron transport (without this, photosynthesis doesn't happen). In summer, when light levels are often high enough to destroy photosynthetic apparatus – a problem that is exacerbated by desiccation and heat stress – the pigment levels are lowered to reduce light capture and prevent over-excitation and destruction. Phycobilins, when not required for light capture, are used as a nutrient source instead. Such seasonal adjustments in pigment levels

Creeping Chain Weed, the small moss-like growth on the rocks, with Spiraled Wrack.

have been observed in green, brown and other red seaweeds, as well as in higher plants.

Another example, Creeping Chain Weed, thrives in the upper shore, albeit under special circumstances. It has a dark, almost moss-like appearance, and may be found in cracks and crevices and on north-facing rocks that face away from direct sunlight. The fronds of its upper-shore brown seaweed companions Channel Wrack and Spiraled Wrack also provide shade for this species on the rocks beneath their canopy.

The characteristic middle-shore red seaweed Pepper Dulse is far less restricted in its requirements than Creeping Chain Weed. This turf-forming red also has the ability – within limits – to resynthesise new pigments in the aftermath of UV damage. In summer the ends of the fronds may lose their red colouration, revealing the underlying green of the universal seaweed pigment, chlorophyll a. In extremis even this colouration will fade and the fronds will appear white. Prolonged exposure to strong sunlight may kill the seaweed, but more often than not a period of cloudy weather will see the colour slowly return. Packed with anti-grazing chemicals (terpenoids, for example), Pepper Dulse is unpalatable to many herbivores. It derives its name from its distinctly peppery taste, and is dried and sold as a spice in Scotland. Nibble a bit the next time you have suitable access (but check the ambient water quality just to be on the safe side).

Seaweeds: the banquet that never was

Pepper Dulse in an unbleached state.

Pepper Dulse bleached by the summer sun.

TOP: An assemblage of lower-shore reds (and browns) from an exposed rocky shore.

BOTTOM: Banded pincer weed species (left); Juicy Whorl Weed (right).

The lower shore and shallow sublittoral are where red seaweeds really thrive. A wander around in this zone might reveal a great number of finely branched, absolutely beautiful specimens, and it is well worth having a look at their structure with a hand lens, as they are arguably by far the most beautiful of the three seaweed groups. Common and often abundant lower-shore species include Dulse, Grape Pip Weed *Mastocarpus stellatus*, Irish Moss (or Carragheen), banded pincer weeds *Ceramium* spp. and siphon weeds *Polysiphonia* spp. Many of these typical lower-shore weeds may be found much higher up the shore if local conditions allow, on north-facing shores or those overhung by a woodland canopy, for example. Rock pools may also change distributions significantly.

Grape Pip Weed is another example of a seaweed that varies in appearance, depending on which stage of its life cycle it is in. Red seaweed life histories are even more complex than those of the browns (see examples on pp. 160–161), so forgive us if we gloss over a few details. In Grape Pip Weed, the characteristic seaweed that we see is the gametophyte phase, whereas another (tetrasporophyte) phase of the life cycle appears as a deep red, almost black crust on the rock surface, often at the base of the gametophyte. The crust was originally named as a separate species, *Petrocelis cruenta*, until the relationship between them was described; nowadays the crust is referred to as the 'Petrocelis phase'. The gametophyte is a fast-growing annual, whereas the crust grows slowly and may persist for decades; it seems these two forms may allow Grape Pip Weed to persist under different grazing regimes. In experiments where grazers were removed from an area, the crustose form soon disappeared under the growth of diatoms and macro-algal young stages whereas the gametophyte form produced more fronds and thrived. When both life phases were exposed to grazers, the crustose form flourished as it had its entire surface cleaned of epiphytic growth, but the gametophyte suffered as grazers devoured any new growth. It seems these two forms are a kind of bet-hedging strategy that enables Grape Pip Weed to survive in one form or another under different amounts of grazing pressure (Little *et al.* 2009).

We have previously mentioned another group of red seaweeds, the coral weeds and pink paint weeds; these chalky reds do well in pools, as they are only eaten by specialist grazers equipped to cope with their hardened cell walls. They also thrive in the lower shore and are often at their most beautiful in shaded gullies at the base of the rocks, where their colour deepens to a dark pink, or even purple, in the absence of strong sunlight. Their rather pale rock-pool-dwelling cousins look somewhat drab in comparison. As with the Petrocelis phase of Grape Pip Weed mentioned above, the crustose corallines also benefit from grazing activity keeping them free from overgrowth by epiphytes. Direct grazing damage is ameliorated by having the meristem located well below the surface; conceptacles (external cavities containing reproductive cells) are similarly protected.

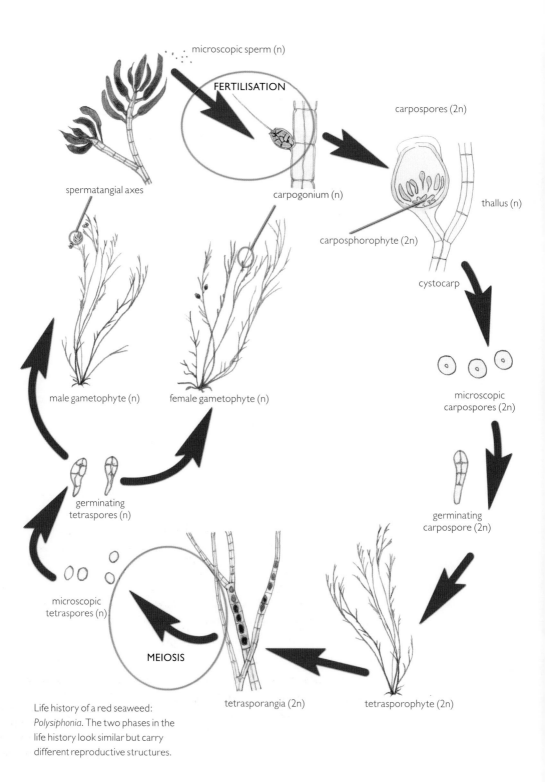

Life history of a red seaweed: *Polysiphonia*. The two phases in the life history look similar but carry different reproductive structures.

Seaweeds: the banquet that never was

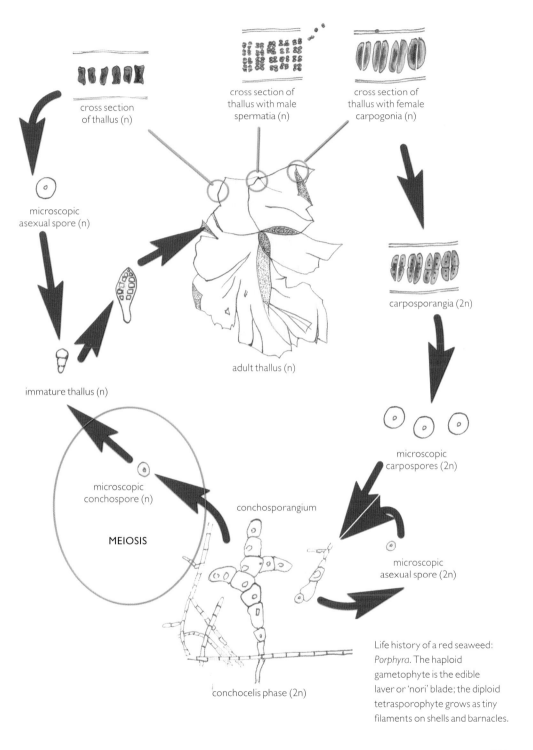

Life history of a red seaweed: *Porphyra*. The haploid gametophyte is the edible laver or 'nori' blade; the diploid tetrasporophyte grows as tiny filaments on shells and barnacles.

Green seaweeds

Green seaweeds are another group with an ancient lineage, dating back a possible billion years; they also share the complex reproduction of the other seaweed groups. There are two main types of green seaweed life history: one where both phases look alike (isomorphic), as in *Ulva* (see p. 163); and another where the recognisable seaweed, the gametophyte, alternates with a microscopic sporophyte phase.

In contrast to the red and brown seaweeds, which are almost exclusively marine, green algae are freshwater specialists with only a small number of representatives in marine and brackish waters. Even so, there are approximately 1,000 species of green seaweed worldwide with some 100 representatives around the coasts of Britain, Ireland and the Channel Islands. Greens have the same chlorophyll complement as land plants, and of the three seaweed groups are the most closely related to them. If brown seaweeds are the 'old men of the sea', then the greens are the 'bright young

Various green seaweeds draped over lower-shore rocks, Black Rock, Pembrokeshire.

Seaweeds: the banquet that never was

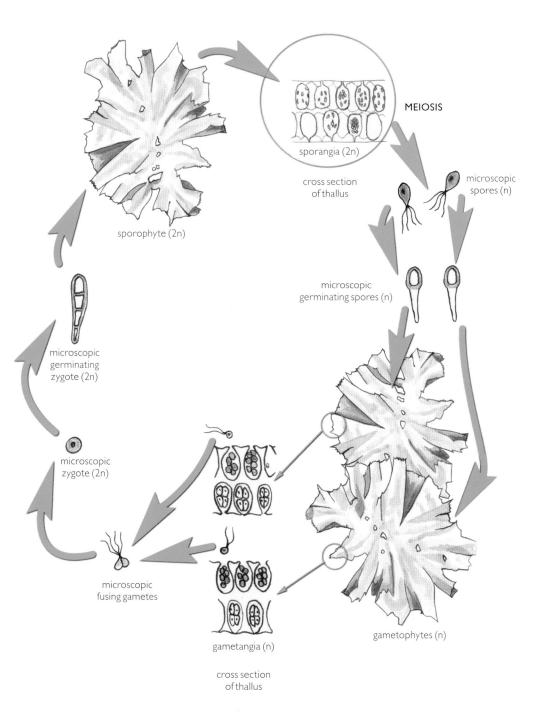

Life history of a green seaweed: *Ulva*.

Rocky Shores

RIGHT: Common Green Branched Weed.

BELOW: Detail of Mossy Feather Weed, a particularly beautiful green seaweed.

things', often appearing on the shore in summer, reproducing and dying within a year. They tolerate wide fluctuations in salinity and are often associated with streams and freshwater runoff on the shore. They are not as rigidly zoned as the browns; indeed, *Ulva* species, often called sea lettuce and gutweed, may be found all over the shore.

Greens are the favoured food of most rocky-shore grazing herbivores, since many of the red and brown seaweeds have unpalatable, anti-grazing chemicals as defence and when mature are too large for most grazers to tackle. It is strange that the most obvious apparent food for herbivores, the carpet of seaweeds present on sheltered shores, is essentially off the menu: the banquet that never was. Red and brown seaweeds' most important function for many rocky-shore invertebrates is structural, rather than edible. Animals can shelter under algal fronds to avoid direct sunlight, a dehydrating wind or the unwanted attention of aerial predators. They can feed on biofilm and epiphytes growing on seaweed surfaces, or indeed be one of those epiphytes. The massive carbon source stored in most shore algae only goes back into the marine system when it is exuded by the plant in life or recycled in death as seaweed fronds break down on the strandline.

Greens are surprisingly difficult to identify with any confidence. The various sea lettuce and gutweed species can only be separated by looking at cell structure under a high-powered microscope. Similarly the green branched weeds *Cladophora* spp. are tricky to distinguish without similar diligence – but do have a look in detail if you have the opportunity, as the branching patterns in this family of green seaweeds are absolutely beautiful. The Common Green Branched Weed *C. rupestris* is dark green and quite wiry, and, as the name suggests, it is the species you are likely to encounter on the middle and lower shore. The slight brown tinge in older specimens comes from a surface coating of diatoms. Recent work on this species' lipid molecules has revealed promising antibacterial properties (Stabili *et al.* 2014).

Mossy Feather Weed *Bryopsis plumosa*, which looks like a diminutive two-dimensional Christmas tree, is another very attractive branched green seaweed. It is usually found in rock pools, where it lends a pleasing green counterpoint to the surrounding pinks of the coral weeds and pink paint weeds.

Seaweeds for food and industry

We started this chapter with evidence that humans have been using seaweeds, nutritionally and medicinally, for thousands of years. Time for a broader look at how useful seaweeds are and may be in the future. With the global human population set to increase for a while yet, there is a requirement for nutritious, sustainable new sources of food – and seaweeds, which are full of flavour and packed with important vitamins and minerals, could be an important part of the solution.

At present there is a cultural divide in how seaweeds are used as a foodstuff. They are routinely eaten and highly valued in east Asia, especially in Japan, China, Korea and countries surrounding the Indian Ocean. In the West the practice is nothing like as popular, although the recent upsurge in the popularity of sushi bars is helping to redress the balance. The red laver *Porphyra* species are the subjects of a multi-billion-dollar industry worldwide. Five species of this genus occur commonly in the British Isles (although the taxonomy is changing rapidly, so take this number with a pinch of salt). Some species are epiphytic on kelp stipes or other seaweed species, others prefer bedrock, boulders, pebbles or even barnacle 'shells', and the genus can occupy all shore heights including the splash zone on exposed shores. In Wales there is a tradition of eating laver as laver bread (it can be enjoyed

---

Ulva seaweed bread twists

Ideal to cook up on the strandline, using dry driftwood to make a small fire.

- Put 1 cup of self-raising flour in a bowl, then add a pinch of salt and a glug of olive oil. Mix in up to ½ cup of torn-up *Ulva* (fresh green weed from a rock pool, well rinsed) – this could be gutweed or sea lettuce. Add slowly to make a good dough.
- Roll into thin sausage shapes. Put one end on top of a stick (if bark is present, peel this off first), and wrap around the stick to hold fast. Toast over the fire. Job done.
- Add a bit of Irish Moss for colour, or Pepper Dulse for extra spice.

Recipe: Clare Cremona

cooked in bacon fat with a fried egg and mashed potato), but worldwide *Porphyra* is better known as 'nori'. Nowadays more than 10 billion sheets of nori are produced each year for Japanese cuisine, but this successful aquaculture industry owes a great debt of thanks to an English algal researcher, Dr Kathleen Mary Drew-Baker (1901–1957). Occasionally there were years when the harvest in Japan failed, and nobody seemed to know why. Dr Drew-Baker's research revealed that laver is another seaweed in which the life-history stages look completely different from each other; the crustose phase, until 1949 mistakenly identified as a separate species, *Conchocelis rosea*, lives on the shells of oysters and clams. This discovery allowed the successful culturing of the species, taking into account all of its life stages, and Dr Drew-Baker is still revered in Japan as the 'Mother of the Sea'. Nori 'fishers' still gather to honour her memory on 14 April each year.

Another, perhaps rather more bizarre, use for seaweeds became popular in Britain and Ireland in Edwardian times. In most large seaside towns, one could take a private 'seaweed bath': a normal bath was filled with hot water and Serrated Wrack, which had previously been steam-treated to facilitate release of minerals, trace elements and polysaccharides into the water. This procedure was supposedly efficacious in the treatment of arthritis, rheumatism, eczema, asthma and stress-related illnesses, as well as encouraging weight loss in the process. These claims are still very much in evidence, and an online search will do much to convince you that Ireland is the place to visit if you wish to try a seaweed bath for yourself.

In industrial terms, most of today's global seaweed harvest is used for gelling agents in food, drinks, cosmetics and pharmaceutical products. The holy trinity of seaweed products for industry comprises alginates, carrageenan and agar, which are all polysaccharides (complex sugars). In the 1880s, Scottish chemist E. C. C. Stanford extracted compounds that he called alginates from British kelp. These chemicals form a stable, viscous gel in water and are therefore invaluable binders, stabilisers and emulsifiers that are now used in toothpaste, soap, ice cream, tinned meat and even fabric printing (alginates thicken the dye without affecting the reaction between the dye and the fabric, resulting in prints that are softer with more vivid colours and sharper edges). Egg Wrack and Forest Kelp in particular are good sources of alginates, which are extracted from the cell walls. Some 16,000 tonnes of Egg Wrack are harvested each year in Ireland and converted to 3,000 tonnes of 'seaweed meal', which is then exported to Scotland for processing.

Oarweed, one of the impressive lower-shore kelp species used industrially.

Carrageenans are produced by red algae and have protein-binding properties, making them ideal stabilisers for cream cheese, ice cream and milk desserts. A small percentage of carrageenan in ice cream can ensure that it melts more slowly, reducing the risk of your treasured seaside treat ending up as a sad mess on the promenade. In Ireland and Scotland, Irish Moss is boiled in milk and strained, and then has sugar, vanilla, whiskey or brandy added to create a jelly-like substance a bit like panna cotta or blancmange, which it is claimed has medicinal properties as well as the inevitable aphrodisiac powers. This versatile seaweed is also used as a clarifying agent in the beer industry, and carrageenans are often added to flour products. Bread, for example, retains moisture better with carrageenans, and the additive has no effect on the action of the yeast or gluten. Elsewhere, there are uses in paints and various cosmetic products (shampoo and toothpaste), and in the printing industry.

Agars are also derived from red seaweeds and can form firm gels, providing a superior alternative to gelatin for vegetarians, even though jellies using agar (or carrageenans) are more opaque, have a coarser texture and less of a melt-in-the-mouth feel. Perhaps the most widely known application for agar is as a growing medium for the cultivation of bacteria in microbiology laboratories.

Fucoidans are another interesting group of polysaccharides, widely found in the cell walls of brown seaweed but not in other algae. With antioxidant, anticoagulant, antiviral and anticancer properties, they can also impede the formation of gastric ulcers by targeting the causative bacterial agent, *Helicobacter pylori*. Fucoidans are found in the highest concentration (up to 20 per cent of its dry weight) in Bladder Wrack, the first alga from which the compound was isolated a century ago.

All of these seaweed polysaccharide compounds have significant medical potential (Holdt & Kraan 2011), and research continues worldwide for medical breakthroughs targeting cancer, boosting immune responses, combating viruses and reducing inflammations, and in the creation of new antibiotics. Recently it has been discovered that a bacterium that lives on the surface of some seaweeds, *Bacillus licheniformis*, may have anti-plaque properties when included as an ingredient in toothpaste.

Kelps have been used on an industrial scale in the production of potash, which, when combined with nitrates, can yield saltpetre for gunpowder. There is potential for many different seaweeds to contribute to the production of biofuels, but the process is complex, fraught with difficulty, and unlikely to happen any time soon.

In agriculture, seaweeds have been used as fodder for domestic livestock for centuries: an alternative English name for Dulse is 'cow weed', and Egg Wrack has been called 'pig food'. In Orkney, North Ronaldsay sheep still graze seaweeds on the foreshore; their salty food source gives the meat a special flavour, and it is sold as a delicacy. Aquaculture also benefits from seaweed use as a sustainable, healthy addition to fish food. It is claimed that salmon fed in this way show greater weight gain, higher levels of omega-3, fewer infections and sea lice, and that the flesh has a better texture, flavour and appearance.

* * *

These last two chapters have concentrated on rocky-shore 'plant' life and its underlying importance to the ecosystem. It is time to turn our attention to the various invertebrate groups that make this intertidal habitat so special.

Stingers, squirts, sponges, mats and worms: the weird and the wonderful

chapter six

No other environment in the British Isles can match the rocky shore for biodiversity, and a visit to a sheltered shore on a low spring tide is likely to provide plenty of material for a student of taxonomy. Beside the well-known groups of plants and animals, there will also be strange, uniquely marine and unfamiliar phyla with few or poorly understood representatives.

It might seem a bit odd to mention golf at this stage in the book, especially as neither of us plays. We know that the zoologist Sir Ray Lankester enjoyed the game, however, because while he was staying at St Andrews, Scotland, in 1885, he named a new species of the phylum Sipunculida, dissecting it between excursions to the Royal and Ancient links: *Golfingia macintoshii*. In his memoirs Lankester explains the derivation of the first half of the name: 'I have accordingly ventured to dedicate the new genus of sipunculid worms indicated by this specimen to the local goddess whose cult is associated with the most ancient of Scottish seats of learning.' The second part of the name honours Professor Mackintosh, who provided the specimen.

Sipunculids (or peanut worms) are worm-like animals that can withdraw the front part of the body into their bottoms. Despite their high taxonomic status of a phylum, there are only 320 known species worldwide. Compared with molluscs or arthropods, they are an obscure group that few people know a great deal about – but such is the enigmatic nature of the marine environment. There are many of these esoteric phyla with small numbers of species, each with distinct characteristics that isolate them from any other major group.

OPPOSITE PAGE:
A syllid worm. The top portion is the adult, which has grown a yellow, sexual phase called a stolon at the end of the abdomen. This will be released into the plankton to reproduce.

Rocky Shores

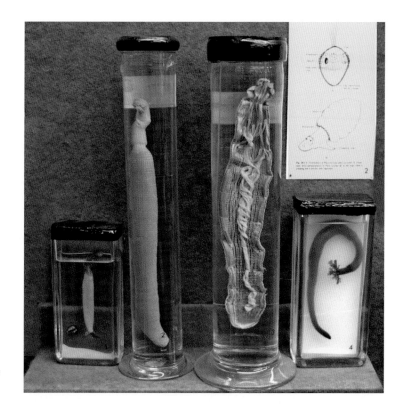

Preserved sipunculids in the University of Edinburgh collections. *Golfingia vulgaris* is the small specimen on the left, collected at Millport. The two large central specimens are of *Sipunculus nudus*.

Strange worms that are not worms

As with the peanut worms, simply calling any long, thin animal a worm can be confusing, and on the rocky shore there are plenty of perplexing possibilities. Nematodes for example are roundworms that often parasitise mammals such as dogs, but here on the shore they are prolific, with free-living examples as well as parasitic forms. Large numbers of these white, unsegmented nematodes can accumulate in kelp holdfasts. Members of another phylum, the Nemertea, are also unsegmented but much more colourful. The common names for this group are proboscis worms or ribbon worms, and the creatures come in a wide range of forms and sizes. The former name relates to a muscular proboscis above the gut that is eversible (turns inside out) to grab prey, while the latter effectively describes their appearance: long and flat. All ribbon worms are active predators that feed on a wide diet of invertebrates. Unlike the nematode phylum, where individuals are less than a centimetre long, the nemertines are more complex and reach lengths up to 10m, as seen in the Bootlace Worm – whose scientific name, *Lineus longissimus*, describes the animal perfectly. We have found them in the lower shore from the Western Isles through Cardigan Bay and down to the Salcombe Estuary. They are often discovered coiled up under stones on sheltered shores where sediment collects; as soon as they are picked up the coil becomes a mass of slime. Trying to unravel the mass is necessary to demonstrate the extreme lengths that they can attain.

Another notable group of simple worms that are more ancient in their ancestry is the flatworms or platyhelminths, which display their more primitive nature by lacking an anus. Known for their parasitic forms such as liver fluke, there are also small, free-living freshwater forms, but on the shore they can be significantly more spectacular. At over 2cm in length with white, yellow and black stripes, the Candy Stripe Flatworm *Prostheceraeus vittatus* outshines the tiny dark forms found in ponds and streams. There are bright red examples, too, and some of the minute varieties in amongst rock-pool fringe seaweeds are

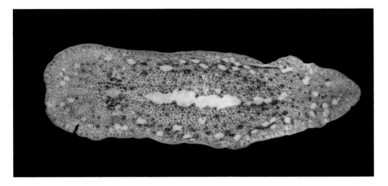

A tiny marine flatworm species (2mm in length) that is usually associated with coral weeds.

A Candy Stripe Flatworm on the underside of a rock on the lower shore.

exquisite. Flatworms glide across the substrate using a combination of cilia (motile, hair-like appendages) and muscle action. Although carnivorous, they will feed on almost anything living or dead, as long as they can surround the prey with their eversible stomach to digest it before bringing this back inside the body. If they are not present in large pools in the middle shore, they will almost certainly be found in the lower shore under rocks or in pools around much of the British Isles.

Primitive life forms and life on the shore

The most primitive group of animals on the shore is the sponges. They appear so inanimate that until the work of Robert Grant early in the 19th century, few believed them to be anything other than a plant. Grant was an anatomist at Edinburgh University but was always looking for material, usually marine invertebrates, to provide students with dissecting experience. In fact, the young Charles Darwin became one of his students in 1827 and often accompanied Grant on his foraging expeditions on the Firth of Forth shores. Grant wrote the first books on sponges and was instrumental in classifying them with the phylum name Porifera, meaning 'pore bearer'. The most commonly encountered sponges are encrusting species on rocks of the middle and lower shore, typically beneath large overhangs of rock or in deep crevices that retain water or face away from sunlight. Sponge cells can be differentiated into various categories, but they do not form organs and are incapable of movement. The matrix that holds these multi-celled creatures together consists of numerous microscopic spicules: three-dimensional spines made of silica or calcium, forming a primitive skeleton. The outer surface has many tiny pores called ostia through which the seawater flows to spaces inside the sponge. An inward current is maintained by a lining of flagellated cells that brings in particulate matter to be filtered from the water. A waste current escapes through larger holes in the surface, the oscula.

An early experiment to demonstrate the primitive nature and loose connection between the cells of sponges involved forcing the creature through a sieve. In more recent times it has been attempted with a liquidiser. The principle of the experiment is to separate the cells from each other into clean, aerated seawater and leave to stand; the cells will gradually reattach and form a multicellular sponge once more. The process is complete in 24 hours, and works best with encrusting species.

Scottish physician and biologist, Robert Grant (1793–1874).

An abundant and widespread encrusting sponge is the aptly named Breadcrumb Sponge *Halichondria panicea*. It varies in colour, ranging from dark green to bright yellow; bright green ones have a symbiotic green alga resident inside them. The encrustation can be quite extensive, and several centimetres thick. Where these large forms grow, there is usually an associated community of crustaceans and other animals that are thought to be attracted by chemical secretions. As in many sponges, reproduction can be by both sexual and asexual means. The former involves releasing a planktonic larva in early summer, while the latter is through fragmentation of the sponge to form new encrustations.

Any thick, rounded orange growths, on the lower shore and occasionally in pools, will probably be Sea Orange *Suberites ficus*, although separating different species is difficult. *Hymeniacidon perlevis*, another orange species, is the most common species across the shore and forms small, asymmetrical encrustations. A very different sponge with a regular shape is the Purse Sponge *Grantia compressa*, the genus being named after Robert Grant. Although this species will attach to rocks in crevices, it is more usually found attached to seaweeds in the lower shore.

All the creatures that we have discussed so far in this chapter are most likely to be found on the lower shore or deeper. They are not adapted to cope with the environmental stresses of the upper shore, remaining

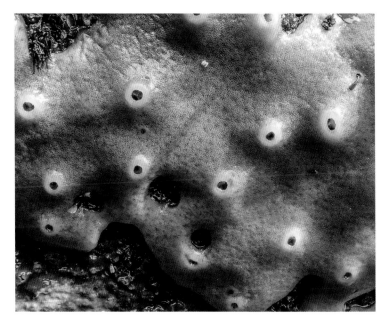

Breadcrumb Sponge, showing the porous surface and the volcano-like oscula through which water and waste are expelled.

The lower half of many rocky shores will typically have an orange sponge, *Hymeniacidon*, encrusting the crevices.

where desiccation is unlikely and the temperature is stable. Scientists have long held the view that life began in the oceans. In fact, ongoing research has put forward rock pools and hot springs as likely places of origin, although the most recent work proposes deep-sea thermal vents, where there is sufficient energy to fuel the many chemical reactions necessary for life. The primitive or lower invertebrates, including the animals described in this chapter and many others besides, are least able to adapt to any changes in the environment and are therefore most at home in seawater, which is remarkably stable both physically and chemically. Despite small variations in a few locations, the sea has a relatively constant salinity. This is particularly significant as the mineral content of the internal fluid and blood of much of life on earth (human blood included) has a similar concentration to that of the sea. In this way a flatworm or sea anemone is living in an isotonic environment where any water loss will equal water gain through osmosis. Some creatures will tolerate the conditions that occur during short periods of emersion on a spring tide, while others avoid them by remaining within humid microclimates, such as under seaweeds.

Another survival necessity for these lower invertebrates is catching food, and we can split them into two main cohorts: filter (suspension) feeders and predators. The former are fixed in place and need to be covered in seawater so they can filter particulate matter, whether organic detritus or microscopic life. Filter feeders make ideal targets for predators such as the flatworms, which lack muscles and can only catch prey items that are not going to struggle. Flatworms can entrap some small prey items in a mucus trail that they release while circulating a rock.

Beautiful stingers

Our next group of the more primitive animals is the cnidarians (phylum Cnidaria), which includes sea anemones, hydroids, corals and jellyfish. Abundant examples can be found on almost any rocky shore in the British Isles. The name is derived from a prominent characteristic of the group: the presence of sting cells (cnidocytes), consisting of capsules (cnidae) that contain threads, which either inject toxin into prey or trap them with a sticky adhesive. All of this impressive preyhunting apparatus is contained within tentacles conveniently located near the mouth. The quarry is usually immobilised by the toxin, but as an additional safeguard the sting acts as a harpoon attached by a thread to the cnidarian body, preventing escape. Some forms dangerous to humans exist in the tropics, including some species of box jellyfish, whose sting can be lethal within 20 minutes. In our temperate region the Portuguese Man O' War *Physalia physalis* is an occasional, rather unwelcome visitor from warmer waters. Contact with this animal's tentacles can be exceptionally painful but is rarely fatal. While it may have the appearance of a jellyfish, this animal is actually a floating colony of hydroids.

Given the primitive composition of these animals, it seems incredible that most of them (with a few exceptions) are carnivorous. After all, the basic form of a cnidarian individual is simply a bag or column, with a wall comprising two layers of cells. The outer cells perform the function of muscles, and can contract and expand to decrease or increase the body size. The mouth at the top of the bag is the only opening. Upon catching prey, the tentacles slowly push it into the mouth and digestion takes place within the bag, where the inner lining secretes digestive enzymes. Waste is ejected back through the mouth.

The body form of the cnidarians varies in size and shape, with anemones being the most obvious on the shore. Dahlia Anemones can grow to 120mm in diameter and are usually brightly coloured.

In contrast, hydroids are tiny and usually in a colonial arrangement where each individual (polyp) is connected to its neighbour by a stem or stolon. Members of the colony usually specialise in feeding, defensive or reproductive roles, and the entire colony attaches to rocks or seaweed in the lower region of the shore to avoid dehydration and the effects of sudden temperature change. The majority of hydroids have a dispersal and sexual phase as a medusa (a tiny jellyfish) that buds from one of the polyps and enters the plankton. *Obelia* species are a widespread and common group of seashore hydroids, with individuals growing on alternate sides of the stolon. *Dynamena pumila* has opposite polyps, and colonies usually grow attached to Serrated Wrack. These two genera are commonly encountered, but a search in the sheltered lower shore under rocks, in crevices, in pools and on the holdfasts of kelp may reveal a number of other species.

Of the other cnidarian groups, we discussed anemones in some detail in Chapter 3, and we will meet hydroid medusae and jellyfish when we look at the plankton in Chapter 10.

Hydroid colonies of *Obelia* (left), showing alternating polyps, covered in brown benthic diatoms, and *Dynamena* (right), on Serrated Wrack.

Colonies of filter feeders

Darwin's first scientific paper, written while he was training to be a doctor in Edinburgh, was on Hornwrack *Flustra foliacea* in 1827. This species is a member of the phylum Bryozoa (sea-mats). It does not live on the rocky shore, but dead remains can sometimes be found washed up on beaches, particularly after a storm. Unfortunately, they are usually disregarded as old, dead bits of sun-bleached seaweed. In size and shape, Hornwrack is like a small brown seaweed, but white. The entire structure is actually a colony of tiny animals; each individual animal or zooid is barely a millimetre across, and they are rarely solitary. The white substance is a calcareous or chitinous protective cover (cuticle) formed from the secretions of many hundreds of individuals joining together to form a beautiful tracery resembling fine lace over seaweed surfaces (typically kelps and Serrated Wrack on the lower shore) and rocks. Another species, *Membranipora membranacea*, is perhaps the most common bryozoan to be found growing on the fronds and holdfasts of seaweed.

At a casual glance under the hand lens bryozoans can be confused with colonies of hydroids, but their detailed anatomy is quite different, and they are in fact more complex creatures, with an anus and proper gut within the cuticle. Hydroids have tentacles with sting cells for feeding, but bryozoans have a lophophore consisting of a crown of tentacles covered in cilia for trapping bacteria and diatoms from the plankton, which they periodically discard and re-grow. Watching bryozoans feeding is captivating: soon after they are submerged in seawater the tentacles come out and wave around before suddenly being withdrawn back inside their cuticular box, only to emerge again shortly afterwards. Like hydroids, bryozoans exhibit a division of labour so that there are zooids for feeding, protection and reproduction.

Bryozoans display amazing variation in their shape, and colonies of *Bowerbankia* are particularly interesting to watch under a low-power microscope as they feed. They are quite widespread around Britain, and in west Wales they are common on the coral weed *Corallina officinalis* in rock pools; the zooids can be seen as little translucent tubes standing erect like vases along the length of the frond. Older individuals are the largest, and may show slight brown discolouration. They have a long, narrow lophophore that emerges, produces a quick wave and is then rapidly withdrawn. A close inspection of some

Rocky Shores

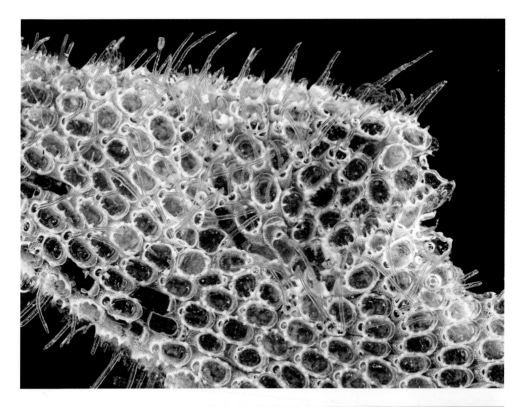

A close-up of the surface of the bryozoan *Electra pillosa*.

Phoronid worms *Phoronis hippocrepia*. Until recently these animals were included with the bryozoans, but the phoronids are now given their own separate phylum.

Stingers, squirts, sponges, mats and worms: the weird and the wonderful

A colony of Light-bulb Ascidians.

seaweed fronds may reveal empty vases, indicating where predators such as sea-slugs have been feeding. These are not the only creatures that eat bryozoans; they are a widely sourced prey for many small predators, including the sea-spiders.

Colonial life seems to be the result of convergent evolution in marine organisms, as there are several groups with sessile (sedentary) zooids forming a tight grouping within a matrix. Not all members of the subphylum Tunicata have these colonies, but those that do not are also filter (suspension) feeders typical of the lower shore. The archetypal rocky-shore tunicate or sea-squirt is a leathery or translucent bag, which contains the viscera. There are two openings at the apex, one for drawing water into the body (inhalant siphon) and one for expelling water and waste (exhalant siphon). The larger animals, at least several centimetres in length, are a favourite with students, who like to squeeze them gently to confirm why they are called sea-squirts. Gills inside the body absorb oxygen, and they also filter material from the water using cilia. The Light-bulb Ascidian *Clavelina lepadiformis* is so transparent that the internal workings are quite clear.

A cartoon of the zoologist Sir Ray Lankester, painted by Leslie Ward for *Vanity Fair*, January 1905.

We need to bring Sir Ray Lankester back at this point. He wrote a bizarre and heavy tome of a book at the turn of the 20th century entitled *Science from an Easy Chair*. Several key marine organisms received entire chapters, alongside others on 'Kissing' and 'Laughter' (he is said to have been an excellent after-dinner speaker). In a chapter called 'The strange history of the tadpoles of the sea', Lankester compared detailed drawings of frog tadpoles and tunicate larvae; superficially they look very similar, just very different in size. Sea-squirts have a primitive appearance, and yet they are chordates (members of the phylum Chordata), and so are close cousins of the vertebrates, including fish and mammals. Like many other biologists at the time, Lankester showed that the larval stage has some chordate characteristics, primarily the cartilaginous notochord from which the phylum receives its name, and which no doubt helps the larva to swim. Our notochord is fused in with our backbone, but in the tunicates it is a flexible rod running the length of the body. Upon reaching a suitable substrate the larva attaches to it by the mouth, and its tail then slowly disappears along with the notochord as the animal transforms into a sessile adult. Walter Garstang, whom we met briefly in Chapter 1, had many theories on marine organisms with advanced larval forms and apparent primitiveness in the adult. The subject is called paedomorphosis, and it is well explained in the book *The Ancestor's Tale* by Richard Dawkins and Yan Wong (2016).

Finding small 'bags' a centimetre or so long attached to seaweeds or the underside of a rock in the lower shore is a sure sign that you have found a sea-squirt, especially when they jet water at you. There are compound varieties, such as *Aplidium* species that hang down from rocks and overhangs. Shaped a little like a red or yellow mushroom, the fleshy end is a mass of zooids on a stalk. But the jewels to look out for on the lower shore (and they can be quite common) are the star ascidians, *Botryllus schlosseri* and *Botrylloides leachii*. Both have their zooids embedded in a rubber-like enclosure, with the former species arranging them in a star-like circle, while the latter has long oval circles. Each zooid has its own inhalant siphon, but the circular arrangement is around a common, shared exhalant siphon to eradicate waste. The colours of these animals range from blue to green, orange and yellow, and they occur on rocks and large seaweeds.

A colonial way of life will provide some safety in numbers, allowing individuals to share responsibility for protection and help with feeding, and, for a sessile animal, having a mate nearby is very useful. The

Colonies of the star ascidian *Botryllus schlosseri* have the small ovoid animals arranged in circular clusters on the rock.

downside is that once a predator such as a nudibranch or sea-spider has found a colony, it can systematically devour the lot, picking out each zooid until they are all consumed.

True worm diversity

Much of this chapter has been devoted to those obscure groups of organisms that are surprisingly common, but typically small, and so can be easily missed. With a sedentary lifestyle, we have seen that filter or suspension feeding is a common method of ingesting nutrients, unless you can entrap your prey like the cnidarians. Another very diverse group worth mentioning in our intertidal tour of the weird and wonderful is the true worms (phylum Annelida). This is a large and varied group that includes the leeches (Hirudinea), earthworms (Oligochaeta) and bristle-worms or polychaetes (Polychaeta); all are segmented.

Bristle-worms are the annelids most likely to be encountered in the marine environment. They are a spectacularly variable group, including planktonic representatives, and it is no surprise that the taxonomy of this cornucopia of body forms is still what might be

Rocky Shores

A minute bristle-worm taken from a kelp holdfast. They can be abundant here, and are just a few millimetres in length.

described as somewhat fluid. This group is characterised by having segments with paired lateral outgrowths (parapodia) that bear chaetae (bristles). Polychaetes, as the name suggests, sport a great many of these. We have heard of a university professor who named his four daughters after bristle-worm species: *Eulalia*, *Nephtys*, *Nereis* and *Aphrodite*, since renamed *Aphrodita* (the worm, that is). While this has never been verified, it is certainly the case that both the names and the animals themselves are rather lovely! Sadly, only *Eulalia* are found on the rocky shore; the others occur on sand and mud.

For convenience, bristle-worms are often divided up into errant (itinerant) and sedentary forms, although it should be borne in mind that this division has no taxonomic validity.

Errant bristle-worms

One of the errant polychaetes is the Green Leaf Worm *Eulalia clavigera*, which is widespread and commonly seen moving over barnacle-covered rocks and around holdfasts, scavenging for dead and dying barnacles and mussels. Being bright green, it rather stands out, although it is quite thin and only about 100mm long. This is the species you are most likely to see on the rocky shore; other errant bristle-worms, while no less beautiful, tend to be much smaller and live in holdfasts and fringe seaweeds.

As mentioned, all polychaetes have side extensions on their body segments called parapodia. In errant forms they are larger and used for locomotion. Leaf worms are also known as paddle worms, because their parapodia are large and magnificently adorned with broad lateral cirri (the paddles). Among these, the *Phyllodoce* species are arguably the most stunning of all polychaetes, reaching almost half a metre in length with brilliant green cirri and a metallic blue body. Living under rocks in the lower shore, these are voracious hunters, and when suitable prey is found they project the upper gut out of the mouth. This eversible pharynx has chitinous teeth that normally face backwards down the gut, but when the pharynx is extended the teeth point forwards, and grip the prey. With the return of the gut, the prey is pulled inside the body. Antennae, palps and additional cirri on the heart-shaped heads of paddle worms are used to detect prey in the dark. They have eyespots as well, but sight is limited to detecting the amount of light rather than for hunting. When disturbed, all errant worms throw their bodies in a series of S-curves; this allows them to swim through the water, though not for long – with the exception of the very efficient paddle worms.

Another group of polychaetes found under rocks in the lower shore consists of the scale worms, which are related to the *Aphrodita* mentioned previously. They may not be immediately recognisable

A scale worm gliding across a kelp frond partly covered in what is left of a grazed bryozoan colony.

Rocky Shores

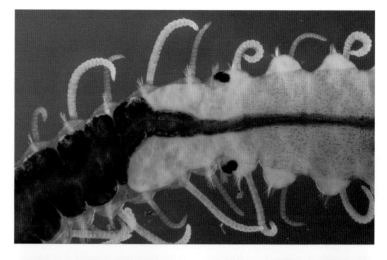

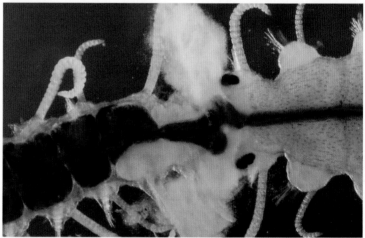

From top to bottom, the brief moment when an adult syllid worm releases a stolon into the plankton.

as worms, for their bodies are short and stout, and covered with scales or elytra. Determining the position and number of pairs of these is important in species identification, and this requires a hand lens. With dense bristles and elytra covering the body, you could be tricked into thinking these animals are rather cuddly, but don't be deceived: they are rapacious predators that eat crustaceans such as amphipods and other polychaetes, using an eversible pharynx like that of the paddle worms.

Certainly, marine worms are a showy bunch of carnivores with a diverse range of both structural and behavioural adaptations. Perhaps the most bizarre forms are found in the syllid worms. Easily overlooked is the small and slender *Eusyllis blomstrandi*, which feeds on hydroids and bryozoans. This species has amazingly long cirri attached to the parapodia, which extend or curl, according to the animal's mood. Females are known to fluoresce with a green bioluminescence, possibly to attract a mate. *Autolytus* species are smaller and thinner still than *Eusyllis*, and they also occur on the lower shore, feeding on hydroids. Their reproduction is similar to that of all syllids. The adult crawls around a rock, feeding on hydroids, and during this time develops a series of new individuals at the end of its body, called stolons. These are the sexual phase, existing as males and females, and after five or so have been produced in a chain they separate and swim up into the water column to become free-living members of the plankton (see Chapter 10). Neither male nor female is able to feed, as the gut is degenerate, but they do have good eyesight, unlike the parents. Each male will perform a spectacular nuptial dance around the female, ensnaring her with mucous strands in which the sperm are located. The female collects some of the sperm and uses it to fertilise her eggs, laying them into a brood pouch which she encircles. She will die after a few days or weeks, and the eggs then hatch into larvae that develop into new adults. This reproductive method is known as 'epitoky', and it is just one of many diverse strategies that polychaetes use to multiply. Some, like the Sand Mason, have even more complex reproduction with several planktonic phases. This is a sediment-shore species, although it also occurs where sand and mud accumulate on the lower shore of sheltered rocky shores and occasionally in rock pools; it is a good example of a filter-feeding worm species.

Sedentary bristle-worms

Sedentary polychaetes are invariably filter feeders. They live in either permanent or temporary tubes, and do not willingly leave them. Parapodia and chaetae hold the body in place, and the head has a crown of tentacles that collect detritus from the water and move it to the mouth by means of cilia. All polychaetes obtain oxygen by diffusion over the entire body, but worms hidden away under stones on sheltered shores may be living in areas of very low oxygen levels. To cope with such conditions, these species have large gill threads that are bright red with the blood. The haemoglobin present in polychaete blood is based on iron like ours (although molecularly different and less efficient). Not only does it collect and transfer oxygen to the tissues, it is also used to store oxygen at a time of plenty for use when levels are low, such as at night.

Terebellid worms such as the *Neoamphitrite* species are typical of filter feeders, and have three separate red gills behind the tentacles. They can be found living amongst large kelp holdfasts on sheltered shores, especially if there is some sand nearby, and can grow substantially, to form a broad, tapering worm some 30cm long. Mucous streams flow along tentacles that act as narrow gutters, filtering out food and dropping it into the mouth at the top of the body. Anything unsuitable is rejected.

The most common sedentary worms guaranteed to be found on almost every shore live in spiral tubes just a few millimetres across. These, the spiral tubeworms *Spirorbis* spp., consist of a number of species, but each tends to associate with a specific substrate. For example, *S. spirorbis* can occur in large numbers on the fronds of Serrated Wrack, *S. corallinae* on coral weed, and *S. inornatus* on red seaweeds, kelp holdfasts and the undersides of the 'buttons' of Thong Weed. The spiral growth form may prove beneficial in coping with the mechanical stresses associated with living on a flexible seaweed substrate. Instead of long flowing tentacles, these tiny worms have a stiffer set of radioles that collect material suspended in the water. Spiral tubeworms are hermaphrodite; they release sperm into the water which is then picked up by their neighbours to fertilise their eggs. These hatch inside the tube and the larvae are released on neap tides, often throughout the season from spring to autumn. They spend very little time in the plankton, probably so as not to leave the zone where the seaweeds they frequent are located. Doing this during a neap tide means that they have a chance to develop a tube sufficient to stop desiccation; a spring tide is more likely to leave them high and dry for longer.

Several other sedentary worms are frequently seen on the rocky shore living in white calcareous tubes, but these are usually glued to the rock. The keelworm *Spirobranchus lamarcki* secretes tubes that have a triangular cross section and can be up to 30mm long. The animal lives among loose stones or rocks, and such tubes are easily damaged on shores when the substrate is moved by the waves. Both spiral tubeworms and keelworms are more commonly encountered towards the bottom of the shore, as their tubes are porous and therefore not as efficient as a mollusc shell at keeping water in. Polychaetes have a planktonic larva that enables them to disperse to new shores and, in the case of the keelworms, find a fresh hard substrate to colonise. Unfortunately, they do not always choose a rock but will foul the undersides of boats and cover shellfish stocks.

The tubeworm found on coral weed, *Spirorbis corallinae*.

* * *

This chapter has examined some of the less well-known but nonetheless common groups of creatures that inhabit the rocky shore. Mostly small, they provide fodder for the more visible animals that we will explore in the following chapters.

Barentsia sp., a stalked entoproct, on a *Dynamena* host. In the marine environment there are several small, easily overlooked phyla, of which Entoprocta is one.

Molluscs: the mantle of respectability

chapter seven

The phylum name Mollusca is derived from the Latin word for soft, but it is the hard, calcareous shell for which many of the group members are renowned. Shells may be in one part and domed (limpets) or spiralled (topshells, periwinkles and Dogwhelks), in two parts (mussels), absent (sea-slugs, octopus) or hidden (sea-hares and cuttlefish). Unlike the arthropod exoskeleton, which has to be shed to allow growth to occur, the molluscan shell can be enlarged as the owner matures, and so the same shell may be a companion for life.

Worldwide, there may be over 75,000 species of mollusc, and representatives are found in terrestrial, freshwater and marine environments. They are one of the most diverse groups of marine organisms and can be found living near abyssal oozes, riding surface oceanic currents, and as abundant and conspicuous members of rocky-shore communities. Some are impressively long-lived: in 2006 a specimen of a bivalve called *Arctica islandica* was discovered that was over 500 years old.

The molluscan body plan varies considerably, as can be seen by comparing squid, cuttlefish, bivalves and chitons. In marine snails, the commonest members of this phylum on British and Irish rocky shores, there is a bilaterally symmetrical, unsegmented body consisting of a head, a muscular foot and a soft visceral mass in which the main body organs are situated. Think slug or garden snail. The visceral mass is covered by the mantle, which is in contact with the inner surface of the shell. The shell is formed on a matrix of protein, called conchiolin, which is stiffened by the addition of calcium carbonate crystals. New shell is secreted from the whole outer surface of the mantle, allowing the existing shell to thicken while new shell is also added at the mantle

OPPOSITE PAGE:
Common Limpets making good use of the available space.

edge. The head contains sensory organs such as tentacles and eyes and the impressive radula, the feeding appendage that is found in most molluscs (apart from the filter-feeding bivalves) and is unique to this group. As mentioned in Chapter 3, limpet radula teeth are made from the strongest biological material known to science. The teeth are set in a ribbon of horny tissue, which looks like a long thin tongue, not unlike that of a butterfly. The ribbon is drawn backwards and forwards across the rocks like a file, rasping material in the process. Detached food is then transported, as if by conveyor belt, to the digestive tract. As teeth are worn out at the cutting end of the ribbon, new ones are produced at the other. Molluscan diets vary, and the structure of the radula has evolved accordingly.

Limpets

The Common Limpet *Patella vulgata*

Limpets are a public relations disaster. One of the shore's most interesting animals is disguised as an unexciting dome of shell that doesn't appear to move or show any signs of animation at all. They can even be mistaken for part of the rock surface. Limpets do in fact move, but mostly at night when the tide is out, under calm seas, or on rainy, humid days when few people visit the shore. To counter the limpet's lacklustre reputation, we used to carry out an investigation with university students at Dale Fort Field Centre that involved three separate visits to the shore. On the first, at low water during the day, ideally when it was dry and sunny, we marked limpets with a number that was replicated on the rock adjacent to the animal. Additionally, a line was drawn down the shell and continued across the rock for approximately a centimetre. The second visit, at low tide at night, was to record whether the numbered individuals had moved from their day-time position, and if so, in which direction and how far. The student groups were equipped with torches for safety while limpet hunting, and the sight of many small beams of light travelling over the shore was slightly surreal. The third and final visit, at low tide next day, was to record whether the marked limpets had returned 'home' again and whether on their return they had orientated themselves such that the line from shell to rock surface was contiguous, demonstrating that the animal had returned to its starting position and was aligned as before.

Limpet marking: stage 1 during the day, when limpets are at home.

The limpets rarely let us down. Most of the animals moved, which caused great excitement among the students, and the record distance covered was a truly impressive 220cm. The vast majority of limpets also found their way home by the following day, even the record breaker (now affectionately called Larry), and not only did they home with great reliability but they also orientated themselves exactly as before, presenting an unbroken line from shell to rock. This homing behaviour in the Common Limpet is well documented and has obvious survival advantages. Limpets living in hard-rock areas will secrete new shell to fit the contours of the surrounding rock. This gives the animals a good seal with the rock surface, enabling them to cling on tightly to keep water in and predators out, but only if they return to the same spot. By contrast, limpets living in soft-rock areas twist around to erode the surrounding rock to fit their shells; in areas of chalk rock, scars can sometimes be seen where limpets have died, revealing their old homes.

A limpet with its shell edge modified to fit the contours of its hard-rock home.

There are locations where the rock surface is reasonably flat – for example in the Bristol Channel, courtesy of the Jurassic Tea-Green Marl. Here, the importance of rigid homing is reduced (one unoccupied bit of flat rock is as good as another), and it seems limpets go in for holiday homes. They may use two or three different areas as refuges on a rotational basis.

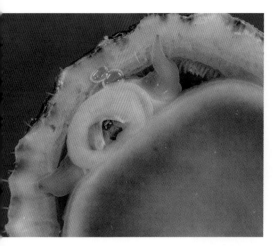

Underside of a limpet, showing its muscular foot, mouth (with radula visible in the middle), head tentacles, mantle and mantle tentacles. The cream mantle tentacles and yellow/orange foot identify this animal as a China Limpet.

So limpets do move, and they can find their way back home with great reliability. As to how they relocate their starting point, there are a number of theories. One is that they simply follow the polarised (directional) mucus trail that is produced as the animal slithers over the rock surface on its muscular foot. Unfortunately for proponents of this theory, research has shown that limpets routinely take a different route back to the home scar than the one chosen for the outward leg of the journey, often using a short cut. Experiments in which the trail is washed off the rock, or the rock surface is remodelled with a chisel, fail to deter the itinerants from returning home. Some researchers suggest that limpets have a 'micro-topographical' awareness of their surroundings, in other words they know their local patch, but limpets don't possess a brain as such, merely a small concentration of neurones, so where would they store this memory?

Whatever the mechanism, it makes perfect sense for a limpet to vary its route away from home on a regular basis, as this will expose the animal to fresh supplies of food away from the area that was grazed previously. Moreover, limpets have endogenous (inbuilt) tidal rhythms, and they adapt their feeding forays to the changing tidal cycles. They also behave differently according to the prevailing weather, how high up the shore they live, and of course how hungry they are. So these deceptively simple-looking animals really are very far from humdrum; indeed, there are many aspects of limpet biology that are still being studied.

Detail of limpet radula teeth, with soft tissue removed.

A limpet feeding trail, showing where the radula has scraped biofilm off the rock surface.

We introduced the process of limpet spawning in Chapter 2. Sperm and eggs are released into the water for external fertilisation, and this has to be synchronised so that all the animals in the local population spawn together. Rough autumnal seas and a reduction in water temperature to the critical 11°C mark are required, although the exact temperature may vary with geographic location. Once fertilised, the egg develops into a larva that joins the plankton for a couple of weeks. When the young animal has grown a shell approximately 0.2mm long, it settles on a rocky shore. As settlement is from the sea and the water covers the lower part of the shore more often than the upper, the majority of young limpets settle on the lower shore. Even if limpets settled all over the shore, most would survive lower down, because thin shells are prone to water loss.

As a limpet grows it begins to need more space, a commodity that is restricted on the lower shore where the favourable abiotic conditions support a multitude of species. Some observations suggest that as limpets grow they move up shore to avoid intense competition (see Branch's chapter in Moore & Seed 1985). They acquire thicker shells as they develop, which helps them with the rigours of dehydration. But whereas some studies have confirmed limpet migration up shore, others have not. On some shores there is an obvious increase in limpet size with height, which might well support a migration hypothesis, but on others the trend is much less clear-cut. And if migration does occur, how does this square with homing behaviour? It is possible that a limpet keeps returning to its home scar until space becomes limiting; maybe barnacle growth nearby becomes an issue. Then there would be a stimulus to move upward to where there should be more space on the rock, and a new home scar could be established. Limpets may reach 16 years of age under favourable conditions, such as on sheltered shores, and this would give them plenty of opportunity to move house occasionally.

When young limpets settle on the shore they are neither male nor female. After a year or two all of them become males, then after four to seven years approximately 34 per cent of the population become females. This particular order of sexual events means that limpets are called protandric hermaphrodites (male first; the opposite is seen in some fish, which are protogynous hermaphrodites, or in other words, female first). The explanation for this curious state of affairs is that when you are very young it is a good thing to put all your energy into growth, rather than reproduction. When you are a bit older, and still growing, sperm is easier to make (energetically speaking) than eggs. The latter can be made when growth is less important, although limpets do grow throughout their lives.

Limpets are infamous for their ability to cling on to rocks. They have a large muscular foot, which is the same shape and nearly as big as the basal area of the shell. Initially it was thought that muscle power and suction, courtesy of the vacuum created between foot and rock surface, were enough to explain their tenacity. The truth is even more impressive. Limpets possess a mucus compound on their foot that can undergo a reversible state-change from lubricant, when the animal is in motion, to 'super-glue', when adhesion is required. This can happen rapidly, as enzymes react with chains of mucopolysaccharides to adapt the function of the mucus accordingly. Attempting to remove a limpet from the rock surface almost always results in catastrophic damage to the animal's musculature and effectively kills it: this is not recommended.

If the foot of a limpet is perfectly adapted for adhesion, then the dome-like shell shape presents the ideal profile for coping with water movement and wave action. The exact shape of the shell, however, varies with location. On some shores limpets get taller, in relation to their basal area, as you move up the shore. Limpets at the top of the shore spend more of their lives with the shell clamped tight to the rock surface to keep in moisture. Clenched muscles cause the mantle to warp, such that the shell it secretes is at a steep angle to the rock surface. Additionally it is likely that upper-shore limpets are less well nourished, with the result that their tissues do not press against the shell (as a well-fed limpet's would), and this again would lead to a steep shell profile. Following this logic might lead us to suppose that limpets on exposed rocky shores should also be tall, as they must have to cling on tightly to survive the extra wave action; there are also greater numbers of limpets on exposed shores, and there may well be less food (in total and per limpet), so they are likely to be less well-nourished too. A 'pointy' profile might indeed be beneficial if the apex of the shell is pointed towards the prevailing wave direction, but it makes no sense if the animal is broadside on to the full force of the waves. In this instance it would be better to be flattened onto the rock surface – so a lower profile is more typical of limpets on exposed shores. The situation is complicated by the fact that shell shape also varies with the slope of the rock. In many ways it seems that limpets are supremely adaptable to local conditions, not only on different shores but also within a single shore, as microhabitat varies. Limpets are often different sizes and shapes in and out of rock pools, for example.

Variation in limpet shell shape: an animal from high up on a sheltered shore (above) and low down on an exposed shore (below).

Most seashore creatures respire with the aid of gills, which function optimally when immersed in oxygen-rich liquid. When an animal is emersed, the gills collapse as their weight is no longer supported by the water; the surface area available for gas exchange decreases dramatically, and if this state of affairs persists the animal can suffocate. Many cunning strategies are adopted by rocky-shore invertebrates to overcome this respiratory conundrum. In limpets, the gills are housed in the mantle cavity (the area between the body of the animal and its shell) and during emersion this area remains full of seawater, allowing respiration to continue as long as there is a sufficient gas supply in the retained fluid. Limpets at rest will often raise themselves off the rock surface just enough to allow gas to diffuse in under the lip of their shell, replenishing vital supplies of oxygen. The slightest vibration near these animals will cause them to clamp down tightly to the rock surface again. They may also lift themselves up if temperatures are high. This allows a breeze to pass over their soft tissues and the evaporation of water from the surface to cool them down, although obviously with the concomitant risk of dehydration. In extreme cases, such 'gaping' can be significant, and this is a sign that the animal is suffering severe stress. Limpets in such dire straits are obviously much easier prey for predators such as Oystercatchers. Healthy limpets will use their muscles and 'super-glue' to clamp down on the rocks if a predator threatens, but the reaction to the Common Starfish is rather less passive: rearing up (or 'mushrooming'), they then suddenly smash the shell edge down, 'stomping' onto the offending starfish's nearest arm (Little *et al.* 2009).

Having established, hopefully conclusively, that there is more to limpets than meets the eye, we could leave it at that and move on to other rocky-shore molluscs. But before we do so, it is worth exploring in more detail the role that limpets play in the intertidal community. Because they are one of the dominant grazers on the shore, feeding on biofilm and young seaweeds, they can affect community composition and as such are effectively 'keystone species'. This designation refers to a species whose effect on community structure is far greater than would be predicted from their abundance or biomass alone. If limpets are removed from a shore, experimentally or as a result of an oil spill, for example, one of the first observable changes will be an increase in the abundance of green seaweed. This may be followed by a growth of brown seaweed sporelings, and so on. In short, remove the grazers and seaweeds will dominate on all but the most exposed shores. Barnacles, mussels and limpets may then be prevented from re-colonising the shore by the whiplash action of mature brown seaweeds. Eventually, adult limpets may invade the site again, gaining some protection from dehydration and temperature stress from the adult seaweeds. By grazing the rock surface, these limpets can prevent new seaweed sporelings from establishing and can even damage the seaweed stipes with their radulae. The next rough sea may then remove the weakened adult seaweeds and open up new space for settlement, so the shore returns to its previous animal-dominated state. Cyclical changes in rocky-shore communities may follow a number of different routes, and outcomes are sensitive to the prevailing weather and how exposed or sheltered the shore is. Scenarios like the one described above might be more likely on a shore of intermediate exposure; very sheltered shores will be dominated by seaweed to start with, whereas very exposed ones might enjoy the green flush associated with green algae but then attract grazing limpets before the growth of brown seaweeds can occur. All of this will vary with the shore's slope and aspect.

Long-term changes in limpet abundance on a single shore have rarely been recorded, although a study of a limpet population before, during and 20 years after a major oil spill is described in Chapter 13. For readers interested in a comprehensive overview of 'all things limpet', an encyclopaedic account is provided by Branch (1981), although it is quite heavy-going.

Other limpets

Before leaving limpets, it is worth noting that there are other *Patella* species besides *P. vulgata*. The China Limpet *P. ulyssiponensis*, for instance, was mentioned in Chapter 3 as a rock-pool specialist capable of eating the calcareous pink paint weed. It is also found on the lower shore (excluding the North Sea coast and that of the English Channel east of the Isle of Wight) and favours exposed locations. The Black-footed Limpet *P. intermedia* is restricted to southwest Britain with a northern limit in north Wales, and again it is absent east of the Isle of Wight. It too favours wave-beaten shores and is a middle- and lower-shore inhabitant. In this species the sexes seem to be separate, and spawning occurs in late summer, slightly earlier than in the Common Limpet.

Another limpet worth looking out for is the diminutive but rather beautiful Blue-rayed Limpet *Patella pellucida*, which usually lives on kelp plants. As the name suggests, its almost translucent shell is embellished with bright blue lines, parallel to the long axis. Some Blue-rayed Limpets live on kelp fronds, where they excavate their own little pit as they feed. Others move down onto the stipe as the autumn approaches to avoid being cast adrift when the kelp sheds its fronds or is damaged by winter storms. Bright blue lines might seem counterintuitive in a species living on the surface of kelp plants where individuals would be highly visible to predators, but it seems that this display of colour might act as a deterrent by advertising 'toxicity'. Similar markings occur in two highly poisonous sea-slugs, *Polycera elegans* and *Facelina auriculata*, which have overlapping distributions with Blue-rayed Limpets. This might be an example of Batesian mimicry, where a non-toxic species mimics the

Blue-rayed Limpets ('*pellucida* morph') clustered on a blade of Oarweed.

warning colouration of a toxic one and benefits accordingly. A variant of the Blue-rayed Limpet (a different species in the making?) lives concealed in cavities within kelp holdfasts; only the total destruction of the whole kelp will leave it a homeless vagrant. This holdfast dweller is rather dull in comparison to its frond-dwelling compatriot, and it is possible that it doesn't need bright blue lines to mimic toxicity as it is too well hidden to be threatened by visual predators. The bright blue-rayed forms belong to the '*pellucida* morph'; the holdfast dwellers belong to the '*laevis* morph'.

Dogwhelks

Dogwhelks are snails with shells in the region of 20–35mm in length. Yellow, purple, brown, orange and even striped individuals exist, but typically in the British Isles they are white. All have a groove coming away from the shell opening, towards the base of the shell, which makes identification pleasingly straightforward. In stark contrast to the other molluscs described in this chapter, the Dogwhelk is a voracious carnivore. It is also a choosy one. The acorn barnacle *Semibalanus balanoides* is favoured over the Common Mussel, which in turn ranks

Dogwhelks, showing a small sample of colour variants.

above other barnacle species. In the Isles of Scilly there are shores that are devoid of barnacles and mussels; here, Dogwhelks will eat limpets (especially young ones), topshells and periwinkles instead, although Toothed Topshells and Edible Periwinkles are usually too large to be viable prey. Dogwhelks even indulge in cannibalism when expedient.

Feeding, which involves boring neat little holes in the shell of the chosen victim, entails a radula, but only partially; the main weapon is the delightfully named 'accessory boring organ'. This device secretes an enzyme that softens the shell of the prey, then the radula is brought into play to remove debris, followed by the accessory boring organ which secretes more enzymes, and so on. The neat, round hole reflects the shape of the boring tool. Once the shell is penetrated, the Dogwhelk secretes a narcotic to paralyse its prey. A barnacle thrashing about may not represent too much of a problem, but a limpet is of course capable of rearing up in an attempt to dislodge a Dogwhelk (or indeed any other animal) that tries to climb on top of its shell. After paralysis comes the insertion of the Dogwhelk's proboscis, which secretes digestive enzymes and turns its unfortunate victim's flesh into a nutritious soup. Feeding is a slow affair; it might take a day to eat a barnacle or over a week to digest a mussel (Crothers 1985), and this can

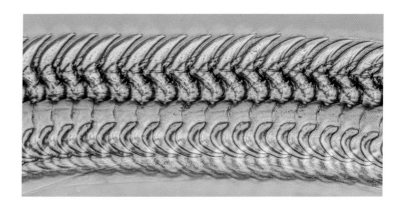

A Dogwhelk radula, showing the impressive array of teeth.

make the Dogwhelk vulnerable to attack. Mussels attach themselves to rocks and each other by sticky byssus threads (imagine a tent's guy ropes and pegs), and they can sometimes use these to trap or even flip over a feeding Dogwhelk, so that it dehydrates or starves to death.

Sexes are separate in Dogwhelks, and animals assemble to copulate in moist microhabitats such as pools and crevices mainly in spring and summer. Aggregations may be well in excess of 30 animals, with females copulating then laying a few egg capsules, then copulating again and so on, as long as food reserves and enthusiasm allow. The yellow, skittle-shaped egg capsules are approximately 10mm high and contain about 600 eggs, but only 6 per cent are fertile. The equivalent of a larval stage is completed within the egg capsule so that when the young Dogwhelks hatch they do so as minute snails, referred to as 'crawlaways'. The infertile eggs within the capsule are their first meal; when this food supply is exhausted they go in search of small prey such as tiny mussels or young barnacles, and so their carnivorous career begins. The shell that the crawlaway starts life with becomes the apex of the larger shell as the animal continues to grow, although it may eventually wear away. Dogwhelks grow comparatively rapidly for the first two or three years of their life until they reach adulthood, then growth ceases. In the unfortunate cases where the animal is infected with the larvae of a trematode, *Parochis acanthus*, growth continues, albeit at a moderate pace, throughout the Dogwhelk's life. Parasitised individuals may be discernible by their larger than average size and additional rows of 'teeth' inside the shell lip. Rapid initial growth makes perfect sense, as young Dogwhelks are particularly vulnerable to crab predation until they reach a 'safe' size. If a Dogwhelk makes it to its third birthday without being eaten, chances are it will survive for at least another three years.

You may have spotted that in the Dogwhelk's life history there is an important difference, compared with many other rocky-shore creatures: the absence of a planktonic phase. Young animals hatch directly onto the same shore as their parents. In studies where animals have been marked on release from their egg capsule, they have been discovered no more than 30cm away one year later. On extremely exposed shores, a Dogwhelk might not move more than a couple of metres in its whole life. Thus, populations are isolated breeding enclaves subject to the environmental factors acting upon that particular stretch of shore. As a result, there is tremendous variability in shell shape and thickness on shores with different wave exposure. A typical sheltered shore supports Dogwhelks whose shells have a long pointed spire and an aperture that is small and narrow, while exposed-shore variants might sport short, squat shells with a wide aperture, and indeed shores of intermediate exposure play host to animals of intermediate shell shape.

If reproductive isolation explains the 'how' of such variety, what explains the 'why'? On sheltered shores, crab predation is a big problem for Dogwhelks. There is a distinct advantage in having a narrow opening to the shell to thwart a crab's attempts to pull the animal out via this route or to gain purchase here to break the shell. It also means there is a smaller area over which to lose water during

Dogwhelks with their egg capsules.

Dogwhelks from sheltered (left) and exposed shores (right). The sheltered forms have a small aperture in relation to shell length and a long, pointed apex, whereas the exposed forms show a larger aperture and a rather squat shell.

emersion. But a small aperture is also disadvantageous because it presupposes a small muscular foot, which is less efficient at keeping its owner anchored in wave-battered conditions. Fortunately on sheltered rocky shores this is rarely an issue, by definition. On exposed shores, what is required is a large aperture, associated with a large muscular foot, to keep the animal firmly anchored during rough weather. For these Dogwhelks, the downsides of having a large aperture, namely an increased area for water loss and increased vulnerability to crab predation, are unimportant because crabs avoid high-energy shores while the extra splash and spray help to offset any desiccation risk. In short, Dogwhelk shell shape and associated foot size equip the animal superbly for life on the particular shore on which it survives. Such characteristics seem to be at least partly inherited, but there is more variation in the juvenile population than in the adults; unsuitable Dogwhelk juveniles are weeded out before they reach maturity.

Dogwhelk predation on barnacles can lead to interesting patterns of distribution on some shores. The Shetland rocky shore (pictured opposite) has a band of acorn barnacles towards the top; just below this there is a zone where Dogwhelks are abundant, and below again there is a band where recruitment of new barnacles from the plankton has occurred. The presumption is that the Dogwhelks have predated the barnacle population heavily at the base of the shore, where conditions are best for the predator. Having exhausted this resource, they have moved up en masse to exploit the higher-level barnacles. The interesting question is, what happens next? As a rule, Dogwhelks cannot survive higher up the shore (but see the one hardy individual ignoring this decree in the picture), as they can only use ammonia as a nitrogenous waste product. Ammonia is easy to make but is toxic and has to be diluted with water before excretion can occur. The higher up the shore an animal lives, the greater the importance of water conservation. Dogwhelks stressed by dehydration die from a build-up of their own waste products, and this places an upper limit on their vertical range. Thus the barnacles at the top of the shore are lucky enough to have settled in a Dogwhelk-free refuge. Since the Dogwhelks can't go up, it follows that they must move down and rely on the prey population having been replenished by young barnacle spat settling out from the plankton. By the time this new resource is exhausted, the upper shore population may have recovered, and so on. If not, Dogwhelk numbers may fall as the predator population has outstripped its resources. There may be a lesson for *Homo sapiens* there if we care to learn it.

As carnivores sitting further up the food chain than grazing herbivores, Dogwhelks suffer from bioaccumulation of certain toxins (hydrocarbons, for example) in coastal waters. A famous – or rather infamous – example of a chemical that Dogwhelks are particularly sensitive to is an antifouling paint called tributyltin (TBT): see Chapter 13.

Purposeful exploitation of Dogwhelk populations by humans for food seems to be rather minimal, quite possibly because they are revolting to eat. In his excellent 1985 article 'Dog-whelks: an introduction to the biology of *Nucella lapillus* (L.)', John Crothers quotes a Major W. W. Ker who offered as his considered opinion that 'to the human palate, even one attuned to "sea food", dog-whelks are distasteful.' Crothers suggests that the unpleasant taste comes from secretions from the hypobranchial gland used in shell penetration. These products have, however, been used by people for other purposes. The Neolithic Cretans (5000 BC), the Minoans (1600 BC) and the ancient Phoenicians crushed whelk shells and extracted various purple dyes from whelk glands. The most famous example is Tyrian purple, extracted from various species of shellfish, which supported a thriving

Dogwhelk distribution on a rocky shore in Shetland.

industry in a number of Mediterranean countries. A smaller-scale operation existed in Ireland, on the island of Inishkea North, County Mayo, in the 7th century, which was based on Dogwhelks; products were exported to exotic locations as far away as Minehead in Somerset. Nowadays there are cheaper methods of producing dyes, and various whelk species are now attracting attention as a potential source of anticancer compounds (Benkendorff *et al.* 2015).

Topshells

Dogwhelks, topshells and periwinkles all have a 'horny' plate, or trap door, called an operculum, that they can close over the shell opening when the animal retreats inside its shell. The operculum gives some protection from predators but is essentially waterproof and helps to prevent dehydration. Topshells are easily distinguished from periwinkles and Dogwhelks as their operculum is perfectly circular, whereas those of periwinkles and Dogwhelks are more teardrop-shaped. Then there is the very attractive mother-of-pearl, or nacreous layer, visible near the shell opening or where the shell has become worn. Other distinctions between these closely related groups of snails include their reproductive behaviour. Periwinkles copulate, whereas topshells release gametes into the water in spring or summer; the stimulus seems to be an *increase* in sea temperature, in contrast to limpet spawning requirements.

A Toothed Topshell (left), with a round operculum, mother-of-pearl and a large white patch on the base of the shell, and an Edible Periwinkle (right), with a teardrop-shaped operculum, no mother-of-pearl and a white columella to the left of the operculum.

A Painted Topshell with its pristine shell.

All British species of topshell have a planktonic young stage called a veliger, which is rather short-lived with the exception of the Painted Topshell *Calliostoma zizyphinum*; this species lays a yellowish ribbon of eggs on the substrate right at the very lowest extent of the intertidal, where the adults live, from which young snails emerge about a week later. Life spans vary: the Grey Topshell *Steromphala cineraria* might live for about three years, the Purple (or Flat) Topshell *Steromphala umbilicalis* has an estimated span of eight or more years, and the Toothed (Common) Topshell *Phorcus lineatus* may achieve 15 years with a favourable wind in its sails. Careful observation of the shell will reveal annual growth checks (a bit like growth rings in trees) that may be used to estimate an individual's age.

The prize for being the cleanest topshell on the shore must go to the Painted Topshell, which retains its shell's pristine condition by behaviour known as 'shell wiping': the animal extends its foot over the shell, which is then wiped with pedal mucus. Shell wiping keeps the shell free from fouling organisms, including algal spores, barnacle larvae and tubeworms. It is thought the nutritional gain from ingesting the material it wipes off the shell might amount to some 20 per cent of the animal's daily requirement.

Rocky Shores

A Purple Topshell (left), showing its purple stripes, and a Toothed Topshell (right) with its large white patch.

Purple Topshells: ventral (left) and dorsal (right) view.

Topshells as a group seem to be particularly sensitive to temperature. The Toothed Topshell suffered heavy mortalities in the cold winter of 1962/63, particularly at the northern end of its British range in north Wales. This southern species only extends eastwards as far as Dorset, but there is evidence from MarClim data (see Chapter 13) that in recent decades its northern limit may have extended in response to climate change. Work carried out by student groups at Dale Fort Field Centre over many years confirms that there is also a seasonal component to this snail's behaviour: individuals move up shore in the summer and down again in the autumn. Such migrations are stimulated, at least in part, by temperature fluctuations, and this species is relatively tolerant of high temperatures. Toothed, Purple and Grey Topshells subjected to temperature stress in laboratory experiments succumbed to 50 per cent mortality at 45°C, 41°C and 35°C respectively (Newell 1979).

The Purple Topshell has a larger range than the Toothed, extending northwards to Orkney, but it is absent from the eastern side of Britain and uncommon on the English Channel coast to the east of the Isle of Wight – yet more evidence of a species range extending with the warming of coastal waters. The Grey Topshell, a lower-shore species, is one of a few rocky-shore invertebrates whose range extends all the way around the coastline of the British Isles (in suitable habitat). However, it seems the most intolerant of raised temperatures and in our experience has been less abundant on the shore in recent years. This animal survives at considerable depths sublittorally, as far as 130m below Chart Datum, so it may be that the vertical range has shifted down shore rather than there having been an actual decrease in overall abundance.

Topshells feed on biofilm and must in all likelihood compete for the same resources. Differential temperature tolerances may explain why typically in the summer, on Pembrokeshire shores at least, the animals are spread across the shore zones, with the Toothed Topshells occupying the upper shore, Purple Topshells occupying the middle shore and Grey Topshells doing best on the lower shore. Differences in the summer vertical ranges, when energy requirements are at their greatest for growth and reproductive output, may allow these very similar species to coexist on the shore successfully without excessive interspecific competition. Of course it is precisely such competition in the past that may have led to the present-day observed distributions.

Grey Topshells have many, thin grey stripes on a cream background.

Periwinkles – when is a species not a species?

Rough periwinkles

Rough periwinkles can be found in most of the shore zones and, remarkably, they can also survive in the lower portion of the splash zone. The presence of a waterproof shell is not enough on its own to enable splash-zone existence. Closing the operculum over the shell opening also helps, but since rough periwinkles spend a considerable amount of time out of water they need a means of maintaining their ecological position when the operculum is closed. To achieve this they use mucus to stick their shells to the rock so that they can close the operculum without falling off and embarrassing themselves. The mucus allows the passage of gas across it to enable respiration but is still waterproof.

As previously mentioned, most rocky-shore animals use gills for respiration, which function best when immersed. Some high-shore rough periwinkles may spend well over 90 per cent of their lives out of water, and so they need an alternative solution. They do have gills but these are rather small and ineffectual. There is, however, also an air-filled space between the body of the snail and the shell, the so-called mantle cavity, and oxygen diffuses from this area into the surrounding tissue, with the shell cavity acting as a kind of lung. If rough periwinkles are kept permanently under water they may even drown; animals in an aquarium tank will always crawl out of the water and sit in the open air when given the opportunity.

Rough periwinkles, some showing the conspicuous grooves that give them their name. Colour variation is large within the group and not always optimal for camouflage!

Even the means of waste disposal are fine-tuned to an upper-shore existence. As discussed above, Dogwhelks use ammonia as a nitrogenous waste product, and so need water to flush the toxic chemical from the body. Water conservation is even more paramount for a splash-zone creature, so rough periwinkles, among others, have evolved the ability to excrete uric acid instead. Uric acid is more complex and requires a substantial adaptation in metabolism to produce, but it is insoluble and so is excreted as a solid, saving precious water.

There are some upsides for an animal that can live this high up the shore. Rough periwinkles eat lichens and biofilm, of which there is a plentiful supply in the upper shore and splash zone; there are also fewer other species competing for this food supply. Battling harsh abiotic conditions does have its rewards.

As you may have noticed, we have until this point failed to supply rough periwinkles with a scientific name. There is a good reason for this: the taxonomy of this group has proved troublesome. At one stage, in the 1980s, there were five species within the rough periwinkle group; this then became four, and now there are three and an 'ecotype' (a subgroup of a species particularly adapted to a specific set of environmental conditions).

TOP: A rough periwinkle holding onto the rock with mucus.

BOTTOM: *Littorina compressa*, one of the rough periwinkle aggregate. The black lines at the base of the rounded ridges help with identification.

Littorina compressa, which used to be called *L. nigrolineata*, is relatively straightforward. It is essentially a mid-shore species, often found in rock pools, and has a yellowish-green shell with black lines in between the rounded ridges. Females lay their jelly-like egg masses in crevices and other suitably damp rock areas.

The other two species, *L. saxatilis* and *L. arcana*, can be impossible to tell apart by just looking at the live animals, especially on shores where they coexist: dissection is required so that the male reproductive organs can be inspected. Another key character in the separation of the two species hinges on what the female does with her eggs. *L. arcana* eggs are laid as per *L. compressa* above, but *L. saxatilis* retains the eggs in a brood pouch, within the mantle cavity. This must be a very good way of protecting young from the rigours of dehydration in this high-shore animal. In all three species the young emerge as crawlaways. Consequently, as in the Dogwhelk, there is isolation amongst breeding populations with no planktonic dispersal phase. Why reproductive isolation has led to different species of rough periwinkles but not of Dogwhelks is a bit of a mystery. The latter show plenty of variation in shell shape and size but are still considered to be one species. Are the extreme conditions at the top of the shore the driver behind evolutionary change in rough periwinkles? If you are interested in getting a flavour of the complexity and variety of arguments about this taxonomic group, a 1993 paper by David Reid of the Natural History Museum will provide plenty of food for thought.

Periwinkle taxonomy may be a little esoteric, but this is where the stuff of life is taking place and species are being created. Or are they? What about the ecotype mentioned above? When the aggregate number of species went down from five to four there was still another player in the pack: *L. neglecta*. The characteristic of this little animal is that it is tiny and lives in dead barnacle shells, an ideal sheltered microhabitat on exposed rocky shores. But when the taxonomists investigated this 'species' they discovered that the tiny version could in fact belong to any of the other three species. It was small but still sexually mature not because it was a different taxonomic unit, but because of where it was living; hence it is now recognised as an 'ecotype', not a separate species.

The Small Periwinkle

The Small Periwinkle *Melarhaphe neritoides* is also an upper-shore and splash-zone specialist like the rough periwinkle. Small Periwinkles are easily small enough to hide in dead barnacle shells and minute crevices, and they favour exposed rocky shores. They are recognised as one of the indicator species of exposure used in Ballantine's Exposure Scale, which we discussed in Chapter 2. Unlike rough periwinkles, these animals have planktonic young, so the female must have access to seawater to release her egg capsules, each of which contains a single fertilised egg. Females living in the splash zone have to crawl down shore to the high tide mark to release their young. Small Periwinkles are even more efficient at using uric acid as a waste product than rough periwinkles, and this

Small Periwinkles tucked away in a sheltered microhabitat, with a rough periwinkle on the open rock surface.

very likely explains why they can survive higher up into the splash zone on exposed shores where these two species may be found together. The extended splash zone on exposed shores favours the Small Periwinkle's predilection for feeding on the Black Tar Lichen *Verrucaria maura*. In contrast to the shells of other periwinkles, the slightly softer outer layer of the shell (periostracum) in the Small Periwinkle extends beyond the edge of the calcified shell lip. This provides a flexible rim to the shell opening, which allows a tighter seal to be made with the rock surface to better conserve moisture. This slight difference in anatomy is enough to put this snail in a different genus.

The Edible Periwinkle

Reaching up to 30mm in height, the Edible Periwinkle *Littorina littorea* is the largest of the British and Irish species, and has paid for this privilege by being gathered by the sack-full for human consumption (this still occurs today, in Ireland and Pembrokeshire for example). Edible Periwinkles go in for recycling; after algae have been scraped off the rock, digested and excreted as faecal pellets, bacteria get to work on the discarded remains. Along comes another periwinkle, eats the pellet and the bacteria, releases these as a new pellet, and the cycle begins again. The planktonic phase in this animal is quite a long one; the veliger larvae spend up to six weeks afloat, having already taken almost a week to hatch from their egg capsule. Young may grow up far away from their progenitors. Even with such a

Edible Periwinkles (alongside limpets and barnacles for scale).

generous period at sea there are parts of Britain that have few (the Isles of Scilly) or no (Lundy Island) representatives of this species, thanks to non-cooperative tidal streams and currents. Otherwise, Edible Periwinkles are widespread and found around all mainland coasts and even into estuaries and on mudflats.

Flat periwinkles

All of the periwinkle species described so far have pointed apices to their shells. Flat periwinkles are easy to distinguish from the others, as they have an almost flat top to the shell, hence the name. This is another aggregate species but this time with only two members, *Littorina obtusata* and *L. fabalis* (*L. mariae* as was). Both species lay fertilised eggs in jelly masses on the seaweeds on which they live; mainly Bladder Wrack and Egg Wrack for *L. obtusata* and Serrated Wrack for *L. fabalis*. Young snails hatch from the egg masses as crawlaways, as we saw for the Dogwhelk and rough periwinkles.

In both species of flat periwinkle, colours vary from bright yellow, through orange and green to a deep brown, and researchers still argue about the significance of this. The following is one possible interpretation, but bear in mind, as with all these golden rules, that it won't always apply. *L. obtusata*, which favours Bladder and Egg Wrack, is often found as an olive-green morph that blends in beautifully with the colour of the brown seaweed on which it lives. With its rounded shell, it even mimics the gas bladders of Bladder Wrack. Bladder and Egg Wrack are middle-shore species and spend about 50 per cent of their time collapsed onto the rock surface during emersion, and such mimicry would seem an appropriate strategy to avoid predation by visual predators (birds and fish). By the same argument, it might seem suicidal to favour a bright yellow morph, as *L. fabalis* does, when living on Serrated Wrack. But Serrated Wrack is a lower-shore species and spends a lot of its life submerged; its fronds look yellow from below when the seaweed is floating. From a fish's perspective the yellow snails would be almost invisible against this pale background, so this should enhance survival accordingly.

Another rather interesting distinction between the two flat periwinkle species concerns their lifestyle. *L. obtusata* takes two years to reach sexual maturity and may live for three or more years; by contrast *L. fabalis* has rather a live-hard die-young approach and is sexually mature and may even die within a year. These behaviours actually relate to survival from crab predation. *L. obtusata* lives further up the shore and thus avoids the intense crab activity associated with the lowest levels. It therefore has time to reproduce in a more leisurely fashion. *L. fabalis* on the other hand can succeed on the lower shore because it is an annual. It reproduces rapidly when crabs have migrated offshore and predation is at a minimum. Thus, *L. obtusata* has a spatial refuge from crab predation, while *L. fabalis* has a temporal one.

Flat periwinkles on Bladder Wrack, showing some of the impressive colour variation.

Herbivorous guilds – who eats what?

With the exception of Dogwhelks, all of the marine snails we have discussed are herbivores, and for the most part they use their radulae to eat biofilm, green seaweeds, young brown seaweeds and lichens. Inevitably, however, the story is more complex when we look at the details, and in marine snails there appear to be 'feeding guilds' related to radula structure.

There are 'sweepers' whose radulae sport wide rows of many, small, relatively blunt teeth, which are brushed across the surface of the rocks and macro-algae sweeping up biofilm. Feeding seems to have little effect on the surfaces they graze over. Topshells are members of this guild. In contrast, the 'rakers' have narrower rows of sharper teeth that can dig into the surface of macro-algae, giving them access to seaweed tissue and microscopic organisms. Rakers can also bite off soft macro-algae (*Ulva* spp.), but have no effect on the rock surface. Most periwinkles fall into this category, with the Edible Periwinkle being particularly fond of green algae. Then there are the 'diggers and abraders'. Here, hardened teeth on the radula can abrade the rock surface and scrape off associated macro-algae and crustose coralline algae that coat the floor of many rock pools. Limpets and chitons belong to this guild. Finally, there are the 'biters and cutters' who bite off chunks of tough macro-algae (fucoids/kelps) or even calcareous algae. The flat periwinkle *L. obtusata* belongs to this last guild, which also includes some species of fish and crabs.

Despite these nuances, there must be some competition for food amongst the rocky-shore herbivores, especially on exposed rocky shores supporting high densities of limpets, all of which seem to be reliant on the seemingly invisible biofilm on the rock surface. Specialists like the flat periwinkle *L. obtusata* and Blue-rayed Limpets, which are some of the very few herbivores that can tackle the big brown seaweeds, seem to have a food source almost to themselves and must benefit tremendously.

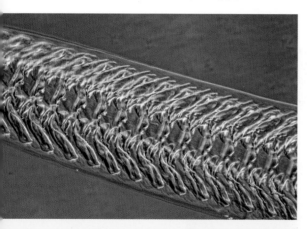

TOP: Radula of an Edible Periwinkle: a 'raker' with narrow rows of sharp teeth for eating seaweed tissue.

BOTTOM: Radula of a Toothed Topshell: a 'sweeper' with wide rows of many small, relatively blunt teeth for micro-algae and biofilm.

Other snails, slugs, cowries and chitons

If the species described so far are the most conspicuous and/or common on the shore, they are very far from being the only molluscan representatives. Many handsome little snails go unnoticed until fronds of seaweed are examined closely. Chink shells *Lacuna* spp., *Rissoa parva*, and the Pheasant Shell *Tricolia pullus*, spring to mind, to name but a few. The Pheasant Shell's convex and bright white operculum is unmistakable. Then there are two species of cowrie that might be found very low on the shore, the European (or Spotted) Cowrie *Trivia monacha* being the most likely, identifiable by its three dark spots on the upper surface of the shell. The other, the Arctic Cowrie *T. arctica*, favours the sublittoral but is an occasional intertidal find. These tiny snails, up to about 13mm in length, are predators on various species of colonial sea-squirt, primarily *Botryllus schlosseri* and *Botrylloides leachii*.

Also sharing the rocky shore are snails without shells: these are sea-slugs, including the colourful nudibranchs. Although they are mainly sublittoral, a good spring tide might yield a species or two. The Common Grey Sea-slug *Aeolidia papillosa* is a good shore

A Common Grey Sea-slug eating a Beadlet Anemone; the nudibranch's egg mass is just below the animal.

example whose diet includes a wide range of sea anemones. In order to photograph it, we once placed a Common Grey Sea-slug in an aquarium tank that happened to contain a Beadlet Anemone. The nudibranch detected the anemone by chemosensory means, shot across the tank (shot being a relative term) and devoured the anemone, then – presumably flushed with success – laid a ribbon of eggs on the nearest bit of seaweed. Most predators won't eat anemones as they have stinging cells (nematocysts), which are used for intraspecific aggression, for paralysing prey and for protection. However, nudibranchs can transport undischarged nematocysts from the ingested anemone through their gut and into storage areas on the end of their cerata (projections on their backs) and use these for their own defence.

The Sea-hare *Aplysia punctata* is a close relative of the nudibranchs but differs in having a delicate internal shell. It is most likely to be seen, sometimes in large numbers, when it moves onto the lower shore to spawn in early spring. This animal can be up to 20cm long, and its colour varies with the type of seaweed in its diet; those eating *Ulva* spp. will be green, while those eating red seaweeds might be almost maroon. Reproduction in Sea-hares is an interesting affair. Simultaneous hermaphrodites, they form a mating chain of several

The Sea-hare gets its name from the head tentacles that look a little like hares' ears.

A chiton, *Acanthochitona crinita*: this species has a broad girdle with tufts of bristles and a fringe of spines.

individuals where each acts as a male to the animal below and a female to the one above. If disturbed while about their normal business they can give off a foul-tasting white slime from one gland and a bright purple fluid from another as a defence mechanism. This species has been recorded from the time of Pliny, in the 1st century AD, and up to the Middle Ages was plagued by an unfortunate reputation attached to the belief that it was poisonous (Fish & Fish 2011).

The curious chitons or 'coat-of-mail shells' are regarded as being the most primitive of the molluscan group, with features that may have a lot in common with the ancestral forms from which all the other types evolved. Their shell is composed of eight articulating plates, which allow the animal to curl up when disturbed, and their appearance is not unlike a flattened woodlouse with no legs. Some species can reach up to 35mm in length. Like limpets, they have a radula with some teeth that are hardened with iron and silicate compounds, enabling them to eat tough seaweeds and encrusting calcareous reds. Their favoured habitat is rock pools and on the undersides of rocks and stones. Of the four families found around our shores, three are well represented by six species. There are two we see regularly: *Lepidochitona cinerea*, a rather smooth light brown to green animal, and *Acanthochitona crinita*, with distinct tufts of bristles surrounding the plates. Close inspection with a hand lens and identification guide will distinguish the various types in the field fairly easily.

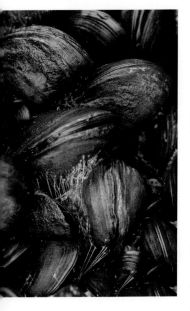

Common Mussels, with byssus threads visible in the centre of the image.

Bivalves

Bivalve molluscs have a shell consisting of two parts, hinged and held together by an elastic ligament. The most common bivalve on rocky shores is the Common Mussel *Mytilus edulis*, which can be found on suitable shores all around the British Isles. Mussels are of course a popular seafood, and there are commercial fisheries based in the Wash, Morecambe Bay, the Menai Straits, Conwy and the west coast of Scotland, for example. A thriving aquaculture industry also exists where mussels are grown on suspended ropes, on stakes interlaced with brushwood, and on poles. These animals are suspension feeders, and will flourish in plankton-rich water, even at salinities as low as 5‰.

Adult mussels attach to the substrate using byssus threads and a mucopolysaccharide glue, but the antics of the planktonic life stage are especially interesting. The veliger larva spends up to four weeks in the ocean currents. Initial settlement, which is temporary and by means of a byssus thread, may be on hydroids or filamentous seaweeds. These young mussels have shells that are less than 0.55mm long and are called 'early plantigrades'; their next move is to drift in the water suspended from a long thread, attach temporarily, drift again, and so on. Drifting on threads has been witnessed in other bivalves and likened to the 'gossamer flight' of spiders – but a much better description is that they are like mini versions of Spiderman. Once attached, they still retain a degree of mobility, and, as noted above, this can afford protection from Dogwhelk predation. If Dogwhelks are present (mussels can detect pheromones in the water), the mussels move together to form clumps and byssus production is increased. Dogwhelks may respond to this defence by favouring the edges of a mussel bed rather than the

Mussels feeding: note the inhalant siphon with the frilly edges that accommodates the gills, and the smooth exhalant siphon.

Coin Shells in amongst a patch of Black Lichen.

danger zone in the middle of a clump. Other predators are starfish, the Common Shore Crab and the Oystercatcher, but mussels are essentially defenceless against these threats. They can even be attacked from the inside: the tiny Pea Crab *Pinnotheres pisum* can live inside the mantle cavity in what amounts to a permanent larder.

Finally, a bivalve that is rather cryptic. The small Coin Shell *Lasaea adansoni* is less than 3mm long as an adult, and lives in upper- and middle-shore crevices, empty barnacle shells, and within tufts of Black Lichen. Unusually for a bivalve, it attaches by a temporary byssus thread so that if the need arises it can crawl away quickly. It is a suspension feeder and filtration rates are rapid, which allows upper-shore individuals to make best use of the limited time they spend submerged. This species can be present in very large numbers on exposed rocky shores but may still elude all but the most observant.

* * *

A visit to a rocky shore will almost certainly yield a variety and abundance of molluscs. Indeed, their absence would indicate that something was very wrong. A lack of echinoderms, on the other hand, is rather par for the course on many British and Irish shores. The occasional coveted discovery of a starfish, brittle-star or urchin causes considerable excitement, and it is to these fascinating animals we turn our attention next.

Echinoderms: animals with tube feet

chapter eight

Marine creatures have a reputation for appearing somewhat unusual. There are animals that look like plants, animals that look like blobs of jelly, animals that look like rock, 'plants' that look like plants but aren't, and so on. Rocky-shore organisms certainly uphold the tradition, but of all of these life forms perhaps the strangest are the echinoderms. This group of exclusively marine spiny-skinned animals includes the starfish, feather-stars, brittle-stars, sea urchins, sea-lilies and sea-cucumbers.

Echinoderm structure

If you were to chop a human being down the middle you would produce two halves that look like mirror images of each other. You could study this wonderful example of bilateral symmetry while waiting for the police to arrive. Echinoderms on the other hand exhibit pentamerous radial symmetry; if you were very small and living in the middle of a Common Starfish you could walk towards the end of any one of its five arms and your route would look identical. Not all starfish have five arms, and in the sea urchins and sea-cucumbers the symmetry is less obvious, but within the animal kingdom this is still a unique, fundamental characteristic of the group. Note that radial symmetry is the norm in plants and that rose flowers, for example, show a beautiful example of pentamerous symmetry.

All echinoderms (phylum Echinodermata) have a skeleton of sorts, made up of a variety of calcareous plates. In the urchins, these plates fuse to give rigid protection to the animal, and in others they are held together by connective tissue to give flexibility, as in starfish. Because the plates are inside the body, covered by a delicate 'skin', this is referred to as an endoskeleton.

OPPOSITE PAGE: Common Starfish en masse for spawning.

Common Sea Urchin: a close-up showing protective spines, some tube feet and pedicellariae with their pincer-like ends.

Spines are characteristic of the whole group, but the starfish and sea urchins have other little structures that stick out of the skin called pedicellariae. These minute pincers have jaws that can snap shut on demand, often located on the end of long stalks. Pedicellariae help with housework, keeping the surface of the animal free from detritus, and deter potential hitchhikers in the plankton such as larvae looking for a substrate on which to grow

With bodies devoid of levers, joints and other associated paraphernalia, echinoderms have had to come up with an ingenious form of locomotion. A whole series of tubes runs through their bodies, containing fluid under pressure. At the surface this system terminates in tube feet that can expand or contract as the pressure in the pipes is adjusted. A valve (madreporite) in contact with seawater allows the adjustment. This hydrostatic arrangement is very powerful; size for size, the arms of a starfish can exert the same pressure as the hydraulic ram of a JCB. As well as locomotion, the tube feet are sometimes also used for feeding and respiration.

Most echinoderms have separate sexes, and fertilisation is external with a series of planktonic larval stages to follow. The larval names vary with the echinoderm group and are rather splendid: for example, sea urchins have echinopluteus larvae, brittle-stars have ophiopluteus larvae, and starfish have bipinnaria larvae (see Chapter 10). As so often, of course, there are exceptions – such as the rock-pool-dwelling cushion-stars discussed in Chapter 3, whose eggs develop into minute starfish without a planktonic phase.

There are approximately 7,000 species of echinoderm worldwide at present. Because echinoderms have a hard skeleton this group is well represented in the fossil record, and many thousands of extinct species have been found. Their ossified skeletons were numerous enough to contribute significantly to limestone formations in the geological record, and that record is extensive: echinoderms have been reliably identified as fossils since the early Cambrian period, about 500 million years ago. Extant British and Irish representatives are rather more modest in number, but approximately seven species of brittle-star, one feather-star, nine starfish, seven sea urchins and six sea-cucumbers might be encountered around the coast on a good low spring tide. This is not the full complement of species found here, but an indication of those you might find with a degree of good fortune.

A side view of a Common Sea Urchin and its tube feet.

Starfish

Common Starfish *Asterias rubens* can certainly live up to the first part of their name. On one occasion, while on a shore survey in 2015 at a site called Monkstone Point in south Pembrokeshire, we were privileged to witness the unusual sight of hundreds feeding merrily on extensive mussel beds at the base of the rocks. They use their powerful tube feet to suck on the bivalve's shell and force the valves apart; a tiny gap, a fraction of a millimetre deep, is enough for the starfish to evert its stomach into and digest the mussel flesh externally. The mussel bed on Monkstone Point was on a thin veneer of mud and gravel over a sandy base. When we revisited the shore for the 2016 survey, storms had removed the precariously sited mussel bed completely. We found just one young Common Starfish in a rock pool.

In the indispensable *Student's Guide to the Seashore*, authors Fish and Fish describe a dense aggregation of Common Starfish on the west coast of England measuring 1.5km long and 15m wide. These animals cleared an area of 50 hectares (3,500–4,500 tonnes) of Common Mussels in four months. Needless to say, starfish are not beloved of shellfish harvesters. Irate fishermen in the past would chop offenders in two and throw them back into the water. Unfortunately this method of predator control was not entirely successful, as starfish are masters at regeneration. Each of the two halves regrew the appropriate missing portion and carried on as

Common Starfish at Monkstone Point. This ephemeral population disappeared a year later, after a storm had destroyed the fragile mussel habitat on which the starfish depended.

before, only now there were twice as many. Although regeneration is a useful survival attribute, sexual reproduction is still the favoured option for maintaining a healthy, diverse gene pool. Since fertilisation is external, it makes perfect sense that Common Starfish release a chemical with their eggs that stimulates others nearby to release gametes as well. Prominent positions are favoured for spawning, such as in the top of a rock or kelp plant.

Mass strandings of all manner of marine species make the news from time to time, and strandings of Common Starfish are no exception. The causes of this phenomenon are hotly debated, and explanations vary with the animals involved, but a recent observation from a team at Plymouth University's Marine Institute may shed light on the occurrence in the Common Starfish (Sheehan & Cousens 2017). Using a drop-down video set-up, individuals were observed with their arms curled under their bodies such that the animal resembled a ball. In a strong tidal flow, the starfish were bowling along the seabed at considerable speed, a behaviour the researchers called 'starballing'. It is unknown whether the animals were at the mercy of the currents or deliberately using the flow as a means of dispersal to new areas, but it could explain a mass-stranding event if the process got out of control in exceptional circumstances.

The Spiny Starfish *Marthasterias glacialis* is restricted to west and southwest coasts of the British Isles and is essentially sublittoral, but it can occasionally be found at extreme low water. This is a large

Echinoderms: animals with tube feet

LEFT: A Spiny Starfish, with a goby for company, on a maerl bed – an unusual individual with six arms.

ABOVE: A close-up of a Spiny Starfish's body surface.

starfish that can reach 70cm across. Animals vary in colour from white to yellow or red and even purple variants, and as the name suggests the upper surface is rather spiny. Its powers of regeneration are so impressive that artificial versions of its growth-regulating compounds (free fatty acids and sterols) are being trialled in cancer treatment (Pereira *et al.* 2014). This animal also gives off saponins, produced as a secondary metabolite, which cause predatory whelks to retreat or even convulse at higher concentrations.

The Northern Starfish *Leptasterias muelleri* looks rather like a young Spiny Starfish but is more obviously spiny and lacks the 'Polo-mint-like' structures around the spines. It is usually purple with paler colouration near the tips of the arms, but shore animals are often green, courtesy of a green alga living symbiotically in the starfish's tissues. This species is another example of a starfish that does not have planktonic larvae; it broods its young on the underside of the body until they are released as crawlaways.

The Bloody Henry *Henricia oculata* might be the most common and colourful starfish around the British Isles; it certainly has the best common name. It is a slightly unusual starfish in that the arms are round in cross section and are rather stiff to the touch. This animal favours current-swept sites and is very likely to be encountered in the coastal rapids on Scotland's west coast. Otherwise it has a wide distribution, especially on the west coasts of Britain and Ireland, and along the English Channel. The Bloody Henry is equipped with a

Rocky Shores

A stunning bright orange Bloody Henry starfish, with Scarlet-and-Gold Cup-corals.

smaller stomach than many predatory starfish, and is a suspension feeder that waves its five arms up into the current, trapping suspended particles in mucus. On occasion it will browse on sponges, hydroids and detritus by the usual (unusual!) stomach-eversion method.

A related species, *Henricia sanguinolenta*, has a northern distribution, particularly on the northeast coast and up to Orkney and Shetland. Its range overlaps with the Bloody Henry's, and they are indistinguishable in the field. Both have blood-red colouration. We have mentioned previously how species classification is changing rapidly in many of the groups of organisms described in this book, especially under the high-resolution lens of molecular genetics. Echinoderm taxonomy is no exception, and *Henricia* is a case in point. There is now talk of a *Henricia* complex of species, of unknown number, that hybridise.

Other starfish such as the Seven-armed Starfish *Luidia ciliaris* and the Common Sunstar *Crossaster papposus* flout the five-arm rule rather significantly. Predictably, the former usually sports seven arms. It is a remarkably fecund species capable of producing over 200 million eggs a year; mortality at the egg and planktonic larval stage must be spectacular, otherwise the seabed would be covered in adults. An unusual feature of this starfish is that the tube feet end in tiny knobs rather than the usual suckers, which may help it to move rapidly enough to pounce on its brittle-star, starfish and urchin prey. The Common Sunstar may have anything from eight to fourteen arms, usually ten to twelve. These and other species of starfish are essentially sublittoral creatures but are well worth looking for at extreme low tides or in deep rock pools.

A Seven-armed Starfish. The prominent fringe of white spines on the arms is helpful for identification, as is the number of arms if they are all present.

The cushion-stars *Asterina gibbosa* and *A. phylactica* have been mentioned before, as they are typically found in rock pools. Both have a rather limited distribution and are restricted to certain west and southwest coasts in the British Isles. Meanwhile the beautiful Red Cushion-star *Porania pulvillus* is found all around these islands, except for North Sea coasts. We have only encountered it in the shallow sublittoral around Arran ourselves, but its bright orange-red colouration should make it easy to spot amongst kelp holdfasts if it strays into the intertidal.

The Common Sunstar is a northern species, rarely seen on the south coast.

Brittle-stars

On the shore the Common Brittle-star *Ophiothrix fragilis* is found in crevices and under stones and seaweed (especially as juveniles among coral weeds), as are other brittle-stars such as the Crevice Brittle-star *Ophiopholis aculeata*, which is more common in the north of Britain, the Small Brittle-star *Amphipholis squamata*, and *Ophiocomina nigra*, which is widespread except for North Sea coasts, from which it is absent south of Northumberland.

In 1984, I (JA-T) was part of a team surveying deep-water habitats off Skomer. At one point I embarked on the deepest dive of my scuba career, to 41m. As my buddy and I descended on a shot line we were acutely aware of the reduction in available light, and I spared a thought for marine plants that attempt to make a living from photosynthesis and how lack of light must enforce an absolute lower limit to their distributions. When we reached the seabed it took a while for our eyes to adjust to the gloom, despite 7m visibility in all directions. The sea floor appeared to be moving! On closer inspection we observed spectacular numbers of Common Brittle-stars; I hesitate to put a figure on the density, but 1,000 individuals per square metre have been observed elsewhere. These animals, in common with all brittle-stars, look like starfish that have been on a strict diet. Five slender arms radiate from a central disc, covered in spines and with tube feet lacking pedicellariae. Feeding habits vary within the group, but the Common Brittle-star is a suspension feeder.

A brittle-star bed at a depth of 40m off Skomer. These creatures can reach densities of around a thousand per square metre.

Just think of the amount of detritus and diatoms (its main diet) that must be required to rain down from surface waters to keep such high densities of these creatures fed.

When exposed to strong currents, Common Brittle-stars often link arms, thereby reducing the chances of being swept away. They have a planktonic larval stage, and young animals have been found aggregating around the inhalant pores of sponges, where they presumably have access to food brought in by the sponge's efforts. Individuals may live for ten years – not quite as impressive as a limpet but not a bad effort for an invertebrate.

Sea urchins

The calcareous plates in sea urchins are fused together to give a rigid outer 'test'. Tube feet are used to pull the animal along, and they must be longer than the spines to be of use. The spines have a defensive function but also contribute to locomotion. Pedicellariae are varied, and present on long stalks; they are perhaps even more vital to slow-moving, essentially solid urchins, that might otherwise become moribund under the load of detritus and unwanted settlers. Urchins scrape food – animals and plant material – from the rock surface using an intriguing appendage called 'Aristotle's lantern', first described by the eponymous Greek philosopher. It consists of a series of plates and muscles, and five protruding chisel-like teeth. As a consequence of this efficient grazing machinery, urchins have been called the sheep of the shallow sublittoral. In excessive numbers they can strip the rocks bare, but under normal circumstances their foraging probably favours high species diversity in the same way that sheep or rabbits can prevent competitive grasses dominating the sward, allowing other, less competitive, herbs and grasses to survive.

The beautiful Common (or Edible) Sea Urchin *Echinus esculentus* comes in a variety of colours from white through to deep purples. Close up, the splendour of the echinoderm's surface detail is a joy to behold. The male and female gonads, or 'roe', from this animal make a valuable food product worldwide. The largest fisheries are in Chile, Japan and the US, but fishery statistics clearly show that present catches are unsustainable. This species is now attracting the attention of Scottish fishermen as well, but fortunately it can be cultured well in aquaria so there is potential for a fully farmed approach to gonad production.

Common Sea Urchin colours vary. This is a particularly handsome individual.

Like many echinoderms, the Common Sea Urchin can occur on the lower shore but does best sublittorally. In contrast, the Shore Sea Urchin *Psammechinus miliaris* lives in crevices on the lower shore. It has a slightly flattened version of the urchin body plan with long, green, purple-tipped spines. Shore Sea Urchins are often found associated with kelp beds, and routinely use bits of seaweed and other miscellaneous shore debris as camouflage.

The Shore Sea Urchin, a spiky animal up to 60mm across, is often camouflaged with bits of seaweed and other debris that it traps amongst its spines.

Echinoderms: animals with tube feet

Many of the rocky-shore species highlighted in this book have a western or southwestern distribution in the British Isles. The Northern Sea Urchin, which goes by the spectacular scientific name of *Strongylocentrotus droebachiensis*, is unusual in being restricted to Scotland, the northeast coast of England and the northern North Sea. A related species, the Pale Sea Urchin *S. pallidus*, can be found around Orkney and Shetland, and needs close inspection of plates and pedicellariae to be distinguished from its cousin. These two are rather like the Purple Sea Urchin *Paracentrotus lividus* in appearance, but their geographic ranges do not overlap, the latter animal being confined to the Channel Islands and the west coast of Ireland. The Purple Sea Urchin is another species unsustainably exploited for its roe, which is prized as a delicacy in Mediterranean countries.

Sea-cucumbers

Sea-cucumbers are elongate, rather worm-like echinoderms with a mouth at one end of the body and an anus at the other. Sightings on the shore are few and far between, and are most likely under stones on the extreme lower shore. The Sea-gherkin *Pawsonia saxicola* can sometimes be found lurking, but at first sight it is rather difficult to believe that this creature is anything other than just a large slug. Its calcareous spines are small and infrequent, hence the soft body and rather slug-like demeanour – but close inspection will reveal rows of tell-tale tube

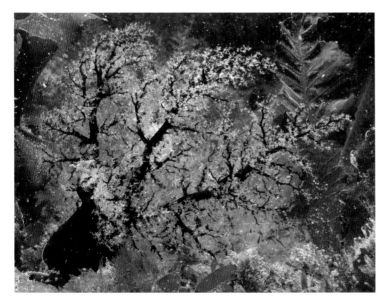

The body of this sea-cucumber is hidden in a crevice, while its large, highly branched tentacles extend out into open water.

feet along the sides of the body, which confirm its echinoderm lineage. The animal uses these lateral appendages to grip onto the rock or sides of the crevice in which it makes its home. Additional tube feet around the mouth, which look like a ring of tentacles, are sticky and trap suspended material before passing it to the mouth. In the highly recommended *Great British Marine Animals* (2005), author Paul Naylor describes this movement as being 'very difficult to watch … without visualizing a child gradually sucking jam from their sticky fingers'.

Sea-cucumber skin is usually thick and tough. It is difficult to comprehend the thought processes involved, but somewhere along the tortuous route of human cultural evolution, somebody thought it would be a good idea to smoke some – possibly because the animal that provided the skin looked like a cigar? They are also eaten, and dried and smoked sea-cucumbers, known as *trepang* or *bêche-de-mer*, are considered a great delicacy in China.

Feather-stars

As indicated at the beginning of Chapter 1, I (JA-T) spent my formative years in Lyme Regis and as a consequence became thoroughly addicted to fossil hunting on the Jurassic Coast between Lyme and Charmouth. This could also be where my interest in tides came from, as it is notoriously easy to get caught out and stranded on this stretch of beach. Regular finds would include belemnites (bullet-like fossils of extinct cephalopod mollusc shells), ammonites (planispiral versions from the same group, related most closely to living *Nautilus* species), fossilised wood and fool's gold (iron pyrites). In fact, my parents and I found enough interesting fossil-bearing rocks over the years to construct a 'fossil-fireplace' from our treasures. However, there was a holy grail to which we aspired: a fossil crinoid otherwise known as a sea-lily. We managed to find a beautiful example preserved in iron pyrites, and I treasure it to this very day – but I was always envious of the perfectly intact specimen in Lyme museum.

To my considerable excitement I discovered in later years that living examples of crinoids still exist. Crinoids are primitive echinoderms permanently attached to the seabed by a stalk (the sea-lilies) or attached only during the early stages of development (the feather-stars). The latter are frequently seen by scuba divers, and occasionally by beachcombers at the level of the lowest spring tides. We have seen the Rosy Feather-star *Antedon bifida* entangled around kelp stipes on a number of shores;

A Rosy Feather-star with a nudibranch *Polycera faeroensis* and Light-bulb Ascidians.

its distribution stretches around most of Britain and Ireland, with the exception of the southern part of the east coast of England.

Feather-stars anchor to the substrate with a 'holdfast' (modified cirri) and wave their arms in the current, trapping plankton and detritus with mucus-covered tube feet. Food particles are then transferred across to rows of cilia (hairs) and down grooves to the mouth, which in feather-stars is on the upper side of the body. The animal appears to have ten arms but in fact there are five, each of which divides into two close to the body. It can crawl over surfaces but can also swim by moving alternate arms up and down. The female keeps her eggs close to her body, where they are fertilised externally by waterborne sperm. Developing embryos are retained in a mucus net, and she holds her arms protectively above them until free-swimming larvae emerge for a brief planktonic phase before settling. It is well that the female is protective, as Corkwing Wrasse *Symphodus melops* seem to have developed a bit of a taste for feather-star embryos.

* * *

Echinoderms are exclusively marine; they rarely venture into even slightly brackish water. Not so our next group of animals, the arthropods. Within this group we do see marine specialists, but many members of the phylum have exploited the land to great advantage. It is time to move on to animals that rely on jointed, lever-bearing appendages for locomotion rather than fluid-filled tubes.

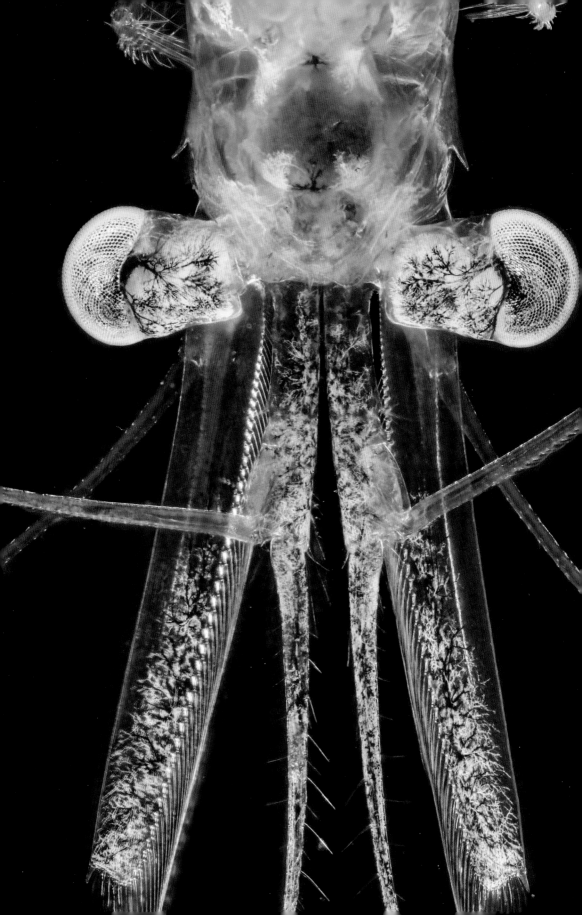

Arthropods: animals with jointed limbs

chapter nine

As anyone who has ever been served lobster in a restaurant will be all too aware, the cuticle of this arthropod is tough, and there are often a number of jointed limbs to cope with as well. These two features – a calcified, semi-rigid exoskeleton and pairs of jointed appendages – are key arthropod characteristics, and indeed 'arthropod' means 'jointed feet'. Crustaceans dominate the arthropod presence on rocky shores, but representatives of many other groups have also adapted to an intertidal way of life.

More than 80 per cent of all known animal species are arthropods, and they surpass all others, both in the sheer numbers of individuals and in the variety of ecological niches they occupy. This phylum of animals includes the Crustacea (lobsters, crabs, shrimps and barnacles, for example), Hexapoda (insects, among others), Cheliceriformes (including spiders, mites and ticks, and sea-spiders), Trilobitomorpha (trilobites, an important fossil group) and Myriapoda (millipedes and centipedes). The taxonomy of this large and varied group has been the subject of considerable discussion, with some authorities proposing that the various subdivisions should be raised to phylum status. This is based on the assertion that the arthropods did not derive from a common ancestor but evolved from separate lineages, so that the group is polyphyletic. Here, however, we stick to the monophyletic approach (assuming all arthropods did have a common ancestor), so we have the phylum Arthropoda, with the crustaceans, insects and so on treated as subphylum groupings.

OPPOSITE PAGE:
Head of an opossum shrimp.

Key characteristics

The arthropod exoskeleton or cuticle is composed of several sublayers, and its main strength is provided by a secretion of a polysaccharide called chitin. In crustaceans (as well as in a few other groups such as the millipedes) this is mineralised, with the chitin impregnated with various calcium salts to create an even stronger skeleton for muscle attachment and protection. The advantages of an exoskeleton are obvious, but there are associated problems as well. As any fully armoured knight of the realm can attest, wearing full battle gear severely restricts growth, movement, and even the ability to feed.

Adapting to so many environmental conditions has led arthropods to develop a wealth of different appendages, from simple legs to complex feeding mouthparts. All of these are segmented, and while some are naked, most display an elaborate array of hairs. To maximise flexibility for walking and swimming, an arthropod's legs are constructed of jointed tubes of varying lengths. Their mouths are also surrounded by a series of small appendages, often looking a bit like fingers holding a knife and fork. Like the limbs, these mouthparts are constructed of small jointed tubes. Typically, there is a set of mandibles with hardened edges for cutting or grinding, then behind these there are other appendages (the maxillae), the structure and function of which vary tremendously to suit a wide variety of feeding methods. When hairs are present they act as a beard to trap particles of food that would otherwise tend to drift away when the animals are feeding on particles in the water.

Like the limbs and mouthparts, the body too is divided into sections. Movement, of body and appendages, is made possible by flexible 'joints'

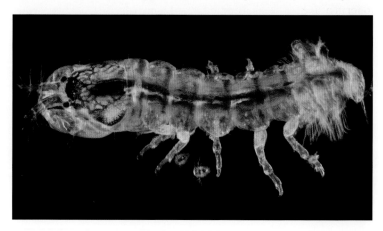

Tanais, a small crustacean easily overlooked on rocky shores. The even smaller creatures swimming in the seawater beside its legs are planktonic nauplius larvae, produced by some crustaceans such as barnacles.

The complex mouthparts of a Velvet Swimming Crab.

connecting the segments. These can be compared to the complex asymmetrical design of the human knee, but what distinguishes the arthropod joints from ours is that there are no gaps within them, just a thinning of the cuticle at that point to allow flexibility. In other words, where our internal skeleton is constructed from many separate bones, an arthropod has a continuous exoskeleton over the entire body. In some areas this cuticle is thick and tough (the segments), and along the edges of these it is very thin (the joints) so that the segments can move relative to each other. These thinner areas also play a part in respiration, as they are partially permeable and allow gas exchange to take place.

Not only does the cuticle provide complete external cover, it extends internally to provide strength and support to structures within the organism. Gills collapse when out of water, but the presence of cuticle within the gills of species such as the Common Shore Crab enables the animal to breathe when the tide recedes. Further diversity in the crustaceans is provided by fusion of upper body plates in some species to form a single carapace covering the segments.

Since arthropods have an exoskeleton, the process of growing can be a challenge. As the body expands, the cuticle cannot grow with it, so it needs to be shed and replaced with a new one. This moulting, or ecdysis, is potentially dangerous for the animal – for not only is it vulnerable during the process, but the old skin may not entirely detach, owing to space restrictions or a host of other issues. An exoskeleton that extends into the body exacerbates the problem, as all of that internal chitinous matter has to be shed as well. Arthropods, crustaceans especially, need to spend up to 90 per cent of the intermoult (the period between moults) preparing for the process. For some time after the moult, about a week for crabs, the animal is so soft that it is unable to feed until the cuticle hardens, so it is important that it prepares during the inter-moult by accumulating food, particularly high-energy oils, in the body for use afterwards. Prior to ecdysis, the haemolymph (internal fluid analogous to human blood) is flooded with calcium salts, some of which will have been recycled from the old cuticle, in readiness for strengthening the new one.

An early study into the moult of Edible Crabs *Cancer pagurus* found that one individual continued to grow for the entire 12 years of its life, regularly moulting every few months for the first five years, after which the frequency reduced to once every two years (Pearson 1908). Lobsters are even longer-lived, possibly up to 50 years, and continue to grow throughout that time.

Insects and others

The overwhelming impression one gets when looking at the earth's various habitats is that the insects dominate the land while the crustaceans excel in the marine environment. While this is largely true, we should not dismiss the importance of the marine insects. Although often overlooked, there are many examples of insects – and for that matter other arthropods such as centipedes and arachnids – living on the rocky shore.

At Dale Fort Field Centre with the windows open on a sunny day, Bristletails *Petrobius maritimus* would dart across the desks and, all too often, inside a computer keyboard – where they probably had a plethora of microorganisms to feast on. The Victorian buildings are constructed such that they appear to emerge seamlessly out of the sea cliffs below, and consequently the offices are treated by seashore

A Bristletail on splash-zone lichen.

animals as part of the upper splash zone. A long yellowish-red myriapod occasionally wandered in too, *Strigamia maritima*, the Marine Centipede. Normally it lives under stones in the upper shore, where it preys on small insects and crustaceans.

For the insects it is often the sheer abundance of a few species rather than species diversity that is important. Larvae of the marine midge *Clunio marinus* are a case in point: huge numbers can be found amongst the seaweed in middle- and upper-shore pools feeding on detritus and diatoms, and they in turn are significant prey items for small predators.

Another insect (though there are arguments as to whether these primitive animals are really insects, and they are often classified as collembolans) is the springtail *Anurida maritima*, probably the most important scavenger on the shore. At low tide, it can be seen climbing over rocks between barnacles and walking on the meniscus of pools almost anywhere on the shore.

Both the midge and the springtail are discussed further in Chapter 3. The Bristletail is appreciably bigger than these two tiny animals, at 12mm in length, and a quick mover to boot. Restricted to higher up the shore, it darts among the rocks feeding on lichen and microorganisms.

Cheliceriformes (arachnids and others)

The Cheliceriformes are an important group of arthropods found commonly on the rocky shore. Spiders may be limited to the terrestrial environment, but other members of the group are widespread and can be found in huge numbers around the coast. Of these, the marine mites are very common, crawling over everything: rocks, seaweed, even other animals. If you take a small pinch of seaweed, say coral weed, and place it in a shallow dish of seawater and look along the surface of the alga for a few minutes (a hand lens is advisable), you will likely find it to be dotted with crawling marine mites. Most belong to one family, the Halacaridae. Some are just visible to the naked eye as oval-shaped, dark or coloured bodies with four pairs of pale legs armed with claws to hang on when waves strike. Under a stereo microscope, we recently watched as a few mites fed on a live colony of the bryozoan *Bowerbankia imbricata*. The colony failed to survive the night, and next day the mites were scavenging inside the dead specimens. These mites are clearly flexible omnivorous feeders. For such an abundant and common group, they are poorly researched and little understood. They would be an ideal study subject for an aspiring student.

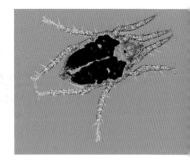

Marine mite, an abundant arachnid found crawling on just about everything on a rocky shore.

The minute *Neobisium maritimum* is the only pseudoscorpion to live on the shore.

On another occasion we found ourselves on an exposed shore in Pembrokeshire staring at a clump of Black Lichen, the tufts of which were just proud of the surrounding barnacles, and appeared to be moving. In the absence of a hand lens we touched the lichen gently, and a couple of tiny invertebrates began to walk across our fingers. More appeared at the top of the lichen and dropped onto our hands, until eventually there were a dozen or so – the greatest number of pseudoscorpions *Neobisium maritimum* we have ever seen in one place. That was in 1999. Between us we had seen only one specimen before then, and probably no more than a further dozen since. This elusive arachnid is just a few millimetres long and remarkably similar to the pseudoscorpions of terrestrial environments. The Black Lichen is a good microhabitat, as it provides both shelter and suitable prey for the pseudoscorpions. In the lab, we have seen them attack Coin Shells, which are of similar size, preferring to start with the muscular foot. Mites are a more common prey item, although pseudoscorpions also feed as scavengers on organic detritus caught up in the lichen.

Sea-spiders are beautiful and frequently overlooked members of the Cheliceriformes. Arachnophobes need not worry, as four decades ago sea-spiders were separated taxonomically into a separate group from the true spiders. They share some peripheral structures, such as the chelicerae or jaws, and look a little spidery – but otherwise they are a unique class of creatures. For one thing they move very slowly, and

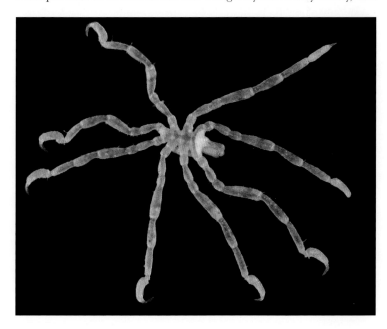

A sea-spider of the genus *Anoplodactylus*.

Arthropods: animals with jointed limbs

A small clump of coral weed, with a marine midge larva (left), a mite (centre) and a cyprid larva (right). Out of focus in the background, a tubeworm.

some are also implausibly thin. Twenty or more species are known around our coast, typically found on seaweeds or sponges. Although they can be as long as 20mm, most are significantly smaller than this, and they tend to be inconspicuous. The easiest way to discover them is to place some fringe seaweeds – coral weed from rock pools or a *Laminaria* holdfast, for instance – into a dish of seawater and watch as sea-spiders such as *Achelia* and *Anoplodactylus* slowly uncurl and crawl out. Given their size, a stereo microscope is an invaluable aid here. The incredibly slender *Nymphon* species are larger, easily visible to the naked eye. Sea-spiders are carnivores and eat a range of sedentary animals including sponges, hydroids and bryozoans, pushing a proboscis into the prey item and sucking out the nutrients. This process is helped by small jaws just inside the mouth.

It is extraordinary that all the sea-spider's organs can fit into such a thin body. The legs dominate here, and it comes as no surprise to find that they have been appropriated for reproduction: for this is where eggs and sperm are located, and the female's genital opening is almost halfway down her third pair of legs. The separate sexes seem to hug each other as they bring their legs together. The male fertilises the eggs as they emerge and then gathers them up to stick them en masse to a specialised appendage, similar to a small leg, near his head. He then looks after the eggs until the larvae hatch.

Crustaceans

As we have been at pains to point out, crustaceans are the arthropods that visitors are most likely to encounter on the shore, and some are impressively large. Surely the aim of any child visiting the rocks is to capture the largest crab he or she can find? No other habitat in Britain or Ireland offers such opportunities to find the largest invertebrates or the most diverse creatures just waiting to be picked up as the tide recedes. At the same time, it is somewhat ironic that the presence of some of the rocky shore's smaller crustaceans can be detected from a kilometre or more away, in the grey/white band of acorn barnacles that typically adorns the middle shore of an exposed shoreline.

When it comes to searching a rocky shore for crustaceans, a bucket, ideally one with clear sides, is useful. Part-fill the bucket with seawater first, so that captured specimens can be dropped in for later examination. The ideal place to search is under large stones and in rock pools (but please put the stones back carefully, and the right way up). Large fucoid seaweeds such as Bladder Wrack and hand-sized stones can be rapidly sluiced backwards and forwards in the bucket to dislodge creatures such as molluscs, which drop to the bottom, and sea-spiders, which descend slowly, waving their legs. You should also see several

Animals found on the shore can be kept for a short time in a large bucket, with a small amount of seawater to ensure oxygenation. Do not place large numbers of crabs with other species, and keep out of direct sunlight.

animals (large and small) actively swimming and darting around the vessel, many of which will be crustaceans that live on, around and under seaweeds and rocks. If your explorations need a focus, you could try investigating the variation in animal communities found on different seaweed species. There are surprising distinctions even between the large wracks such as Bladder Wrack and Egg Wrack, and between the brown and fringe red seaweeds the differences are huge, especially when it comes to tinier animals. The Naturalists' Handbook on *Animals on Seaweed* by Peter Hayward (1988) is an ideal reference.

To separate crustaceans from the other arthropods, have a close look at the front of the body. A willing specimen, say a shrimp or prawn, should be easy to find in a middle-shore rock pool; gently place it in a small glass dish of seawater under a magnifying lens. You might be able to discern three body regions: a head, a middle bit (thorax) and a rear bit (abdomen), with the degree of segmentation differing according to the species. The head has five pairs of appendages. The front two pairs are antennae, and it is only crustaceans that have two pairs of these. Then there are the mandibles and maxillae. The head appendages are involved in the process of eating, and their shape depends on the animal's gastronomic preference, be it carnivorous, suspension feeding or deposit feeding. What happens next as we travel down a crustacean's body varies enormously, and the variations on a theme are only matched in complexity by the bewildering assortment of names given to these jointed limbs. In general terms the thorax appendages might be involved with feeding but are often used for walking or grasping. Abdominal appendages might be conscripted for swimming, or they might be used for carrying eggs or for respiration – either to generate a current past the gills or as the gills themselves.

Blistering barnacles

It was the brilliant 18th-century Swedish naturalist and physician Carl Linnaeus who formally set about the business of taxonomy, or classification of nature, and he went on to classify many new species. The full name of one of the most common barnacles on our shores is *Semibalanus balanoides* (Linnaeus), showing that he was the first to describe it. Embarrassingly for Linnaeus, although he described it well, he classified it as a mollusc, and this remained unchallenged for nearly a century.

The first hint that Linnaeus had made a mistake came in 1830, when John Thompson discovered the barnacle's first larval stage, the nauplius. This has around five moults until it converts to a cyprid larva, the final stage before it becomes an adult. Crucially, a nauplius is a typical crustacean larva, and this meant that barnacle classification needed a complete rethink. The man who took this on, in 1846, was none other than Charles Darwin. He had just got around to looking at some marine specimens from his voyage on HMS *Beagle*, collected in Peru. Over the next 14 months or so he realised what a pickle the classification of barnacles was in, and that a complete re-evaluation of the group was required. Darwin never tackled a subject half-heartedly, so the next seven years of his life were heavily focused on studying barnacles, although he said that this also helped him consolidate his

The yellow individuals among these barnacles are cyprids, the final-stage larvae that settle out of the plankton to attach to the rock and metamorphose into adults.

'transformist' views, which led to the development of the theory of evolution by natural selection.

To Darwin, the similarities between species were the keys to classification. After careful study of the embryology and metamorphosis of the barnacles, he placed them in the Crustacea, subsequently elevating them to subclass status: the Cirripedia. He described many new barnacle species, and in 1851 and 1854 published two volumes entitled *Living Cirripedia*. These are full of information about the acorn barnacles, but an easier modern alternative is Phil Rainbow's 1984 *Introduction to the Biology of British Littoral Barnacles*, which includes a good guide to identification.

Darwin was a prolific letter writer, and frequently wrote to his friend and advocate Thomas Huxley about their shared interest in barnacles. Marine invertebrates were Huxley's passion, and it was he who first referred to barnacles as a 'simple shrimp stuck by its head to a rock covered in a shell'.

So, despite having a somewhat dull and boring appearance, barnacles are actually a compelling enigma studied by many famous naturalists. They also account for numerous records in maritime organisations' accident books detailing grazed legs, damaged backsides and cuts to the arms and hands. After years of climbing on rocky shores, our shins are not a pretty sight. The dangerous beasts responsible are the acorn barnacles, the typical rock-covering biota of the shore.

The barnacle's final, cyprid larval stage (which is easily seen under a microscope) has a good food store of lipid present in the body when it drops out of the plankton. It does not feed, and has the ability to move around the substrate searching for the ideal site to attach to, where it will remain for the rest of its life. When a cyprid larva settles on the rocks it is attracted chemically by the presence of other barnacles of the same species, and this proximity probably helps to provide physical support, as well as making it easier to find a mate and ensure settlement is in an area where adults have grown successfully. Once the larva has settled, the head splits open to reveal a large gland that secretes 'super-glue' to hold the head securely to the surface. With its legs upwards (in air or water), the newly settled barnacle is very vulnerable not just to predation but also to dehydration. It rapidly secretes a series of secure calcareous plates around the body and several above that can be opened, enabling the legs to protrude at high tide. The extended legs (or cirri) are very hairy and kick through the water, netting organic particles, living or dead, which are then brought back inside to be consumed.

Using hairy legs and appendages for filter feeding is a method found across the crustaceans, in contrast to the molluscs and polychaetes, which use cilia (essentially modified hairs). The upper plates also need to open to allow defecation (a process that needs to be timed carefully

A barnacle feeds by rapidly drawing its extended legs through the water to filter small particulate matter.

so as not to bring this discarded matter back with the next sweep of the legs) and reproduction. Most crustaceans need to copulate, and barnacles are no exception; despite being hermaphrodites, they still need to exchange gametes with other individuals. Being stuck in one place could be a dilemma, but acorn barnacles have got around this by having the longest penis of any animal, relative to body size. As if that were not impressive enough, research by Dr Matthew Hoch and his team in 2016 demonstrated that barnacles adapt the shape, length and girth of the organ to best suit the environment they are in.

You would be forgiven for thinking that barnacles are just randomly distributed on any given rocky shore. After all, they look so similar. But further investigation with a hand lens and a suitable identification guide will likely reveal that the species' distributions vary significantly with height above Chart Datum. Wave action also makes a difference: for instance, a pulverising wave limits the growth of one of the common shore barnacles, the slightly softer four-plated *Austrominius modestus*. This species is, however, very tolerant of silt and dilute salinities, so expect to find it in more sheltered conditions and on rocks near estuaries. *Austrominius* is an Australian coloniser and has done

A community of barnacles:
(A) *Chthamalus stellatus*;
(B) *Semibalanus balanoides*;
(C) *Austrominius (Elminius) modestus*;
(D) *C. montagui*.

well for itself all over Europe since the 1940s, when it probably arrived on the underside of shipping.

Particularly prevalent on the shore are the six-plated acorn barnacles *Semibalanus balanoides*, and along with the *Chthamalus* species they create distinct bands in the upper and middle shores. The reasons for this and the intriguing example of competition between them are discussed in Chapter 2. These species have a micropyle, a tiny hole through which they can breathe without opening the plates, which means gaseous exchange occurs with the minimum of water loss. Two further species, *Perforatus perforatus* and *Balanus crenatus*, occur commonly on the lower shore and grow significantly larger than the species higher up, probably because they are able to feed for longer, especially during neap tides when they can filter feed almost continuously. *P. perforatus* does not have the micropyle adaptation and must open the plates to breathe. Consequently it is prone to desiccation, and is limited to the lower shore.

There are further barnacle species to be found in the British Isles. The large Goose Barnacle *Lepas anatifera* arrives on our shores attached to buoys and objects washed in from the open sea. It has a stalk and is quite different to the acorn species. Another rarely encountered species, not often seen as it hides away under rocks, is an asymmetrical barnacle belonging to a separate group with small diversity but a worldwide range. The scientific name, *Verruca stroemia*, gives an indication of what it looks like.

Perhaps the strangest barnacle of all is the bizarre parasitic *Sacculina carcini*. As the species name suggests, it typically parasitises the Common Shore Crab *Carcinus maenas*, and several centuries of global naturalist time have been invested in trying to unravel the complexities of this species. Frenchman Yves Delage produced the first detailed account of the parasite in 1884, but despite many further studies it was not until the 1930s that the full story began to be understood. In outline, the parasite begins like all barnacles, as a cyprid larva. This infects the crab, entering the body to produce an internal root-like structure that pervades the tissues of the host and obtains nutrients from it. Around nine months later a yellow bag develops on the outside of the crab beneath the abdomen, where the female crab holds her eggs. This is called the externa and has a small pore opening to the water through which the nauplius larvae of the parasite are ejected. Originally it was thought that the barnacle was hermaphrodite, but new evidence shows that the first infection is by a

The yellow sac under the abdomen of this Common Shore Crab is the parasitic barnacle *Sacculina*.

female. Once she is established, a male larva arrives and parasitises her to become just a 'gonad' to fertilise her eggs. The effect of all this on the crab host is castration, although the precise nature of the change depends on the crab's age at the time of parasitism. Changes in behaviour also occur – males moving into deeper water, for instance – but the overall problem for the crab is lack of ability to reproduce.

One group of barnacles, the Acrothoracica, are quite literally 'boring'. *Trypetesa lampas* has been found to burrow into the shells of hermit crabs, although the reason for this is still unclear. It has cirri for feeding, but these are slightly reduced in size compared to those of other species. It could be a parasite feeding on the crab's eggs, although it bores into the shells of male as well as female crabs; or perhaps it feeds on surplus food escaping from the crab's mouth. Barnacles still have plenty of secrets.

Copepods: food or fiend?

Copepods hit the scientific headlines in spring 2010 as the world's strongest and fastest animals. Thomas Kiørboe, carrying out research on these animals at the Technical University of Denmark, declared that they can jump at a rate of half a metre a second and are ten times stronger than previously documented. This is all in relation to their size, of course, which is less than a millimetre in length. They are close relatives of the barnacles and are placed in the same class, the Maxillopoda. Barely visible to the naked eye and therefore easily ignored, copepods are among the most abundant animals in the ocean. We discuss them in Chapters 3 and 10, as they stay within the water column, moving in and out with the tide or becoming trapped in rock pools. These tiny individuals are an important food source, linking creatures at the bottom of a food web to the predators above, including other crustaceans and fish.

Copepods, like barnacles, include species that have evolved to become parasites. In fact, it is thought that hardly any group of marine animals has escaped the attentions of these tiny fiends: fish can be especially common hosts to some large and quite extreme parasitic forms; many crustaceans, including shrimps, prawns, mysids, isopods

Arthropods: animals with jointed limbs

Copepods are the most common animals within the permanent marine plankton.

and amphipods, have plenty; and so do starfish and other echinoderms. Hosts are infected by the nauplius larva stage, which may be an internal or an external parasite. Development from the larval state usually follows a course in which the parasite loses its complexity to become a simple degenerate, absorbing nutrients from the host.

Malocostraca: opossums, gribbles, hoppers and crabs

Barnacles and copepods are perhaps the crustaceans that visitors to the shore are least likely to think of, despite being the most abundant. After all, although you might get a nasty graze from the barnacles, you are unlikely to receive a painful nip. The animals most likely to inflict pain of this sort are the Malocostraca, a class containing the crabs, shrimps and prawns, the emblematic crustaceans. Although they look very different, these animals still share many characteristics with the Maxillopoda.

The specially adapted appendages are an important feature of crustaceans, whether for feeding, locomotion or ventilating gills, but the array of appendages, from front to back of the body, varies in structure and function across the range of species. In more primitive types such as the opossum shrimps or mysids, the appendages are mostly uniform in structure across the length of the body, and they lack pincers. These species are believed to belong to an ancient order (Mysida) and are not to be confused with the shrimps and prawns you might keep in the freezer. If there is a slow-moving stream flowing across the seashore, take a shrimping net and sample from the green *Ulva* species (gutweeds

The delicate-looking opossum shrimps have hairy limbs for swimming.

and sea lettuce). Mysids feed on detritus; they prefer dilute seawater and are therefore common in estuaries and lagoons as well as in pools on the rocky shore. They can remain suspended in the water by the beating legs. The ancestral crustaceans were probably planktonic, and mysids have retained the largely undifferentiated limbs and simple locomotion associated with this.

Further down the shore, picking up any sizeable rock on a sheltered stretch is likely to yield many hundreds of crustaceans waiting for the return of the tide. Try not to let the more conspicuous crabs distract you. In the detritus below, small, flat shrimp-like crustaceans will be rapidly flicking their abdomens to escape. These are members of a large order, the Amphipoda. Look carefully at the rock surface and you may see tiny, flattened creatures like woodlice sliding over the surface. These are species of *Jaera*, which belong to the same order as woodlice, the Isopoda. The crabs belong to the order Decapoda, and these three orders contain the most visible mobile crustaceans on the shore, displaying a tremendous range of form and variety.

Isopods

Across the entire shore the diversity of Malocostracans shows an interesting zonation. Life began in the sea, and isopods and amphipods radiated out across most of the niches. Colonising the shore, they gradually adapted to the changeable physical and chemical conditions including the variable salinities of estuaries or where streams cross

Arthropods: animals with jointed limbs

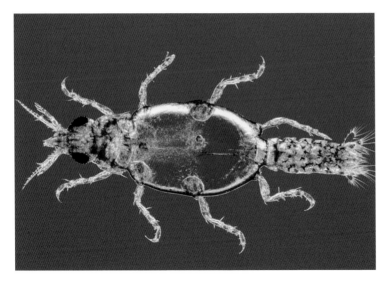

A female *Gnathia maxillaris*. This strange-looking isopod is common in crevices, where she broods up to 100 larvae in her swollen body before releasing them into the plankton.

the rocks of the shore. Isopods went further, adapting to increasing desiccation and moving into terrestrial environments. Most retained gills, demanding damp conditions to enable ventilation, but some woodlice such as the Pill-woodlouse *Armadillidium vulgare* can thrive even on dry chalk downland. In most cases on the shore, the two orders are easily told apart by the isopods being flattened top to bottom and the amphipods compressed laterally. As the latter often lie sideways to the substrate, however, it is important to see where the legs protrude. Neither possesses a carapace like the crabs, but both have a clearly segmented body. Isopods also lack any form of claw or pincer at the front.

The pleasingly named Gribble *Limnoria lignorum* is a species of isopod that bores through wood, including wood pilings and platforms, within the tide zone. Large pieces of driftwood washed onto the shore that have been 'gribbled' will be completely eaten out with burrows and holes. The Gribble is only 1–4mm long, but some isopods can be much larger, and the top of the upper shore is the place to find the biggest; this is the Sea Slater *Ligia oceanica*, the most woodlouse-like of the marine isopods, and it can officially grow to 30mm. On the west coast of Ireland we have seen individuals exceed this size, with some real monsters climbing limestone boulders in daylight without any apparent care for predators (mainly land birds). Normally, Sea Slaters hide by day, emerging at night to feast

The head of a Sea Slater.

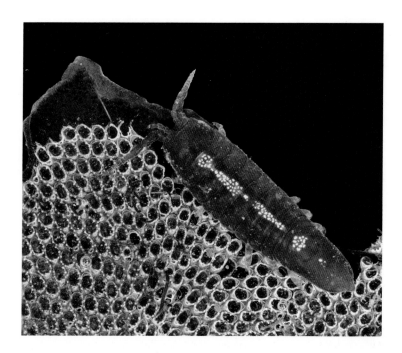

A young *Idotea granulosa* resting near a bryozoan colony.

on anything the tide may have left. They happily scavenge on detritus, especially of animal origin, but when abundant we have seen them eat young amphipods. They will also eat diatoms stuck to rocks.

If you have tried the technique mentioned above of washing seaweeds in a bucket of seawater, you have most likely already come across a common isopod family of the shore, the Idoteidae. These species are rapid swimmers and can be distinguished from Sea Slaters by the greatly elongated last abdominal segment, called the telson, although their habitats are also quite different. Underneath the telson are the gills. Large adults of the common species *Idotea granulosa* occur on fucoid seaweeds; the males grow to nearly 20mm, but the females are much smaller. The tiny juveniles can be found by examining the small delicate fringe weeds of rock pools, such as coral weed. Possibly these are the nurseries for the young stages, which will find plenty of diatoms and detritus here on which to feed.

Any isopods that keep on swimming in your bucket of washed weed after initial disturbance may be something other than *Idotea*. Place one on your hand, and if it curls up in a tight ball you have a member of the family Sphaeromatidae. They live amongst seaweeds and rock crevices and are likely to be *Sphaeroma serratum* or *Dynamene bidentata* (not to be confused with the hydroid genus *Dynamena*). The former, around 10mm long, is a warm-water species and only extends as far north as north Wales, while the latter is distributed well into Scotland. It is also smaller, at around 6mm. The young stages feed intensively all summer on seaweed, storing large quantities of food in preparation for the adult phase, which doesn't feed. During the autumn, a male will take up residence inside old barnacle shells and crevices with a harem of females. After a new brood is produced the females die, leaving the male to look after the young for several months. Astonishingly, he may continue in the absence of food and help produce a second brood the following year.

Amphipods

Although the isopods have radiated out across the shore, adapting to most microhabitats, they have all retained the same essential form, with minimal appendage development; in fact their name means 'equal feet'. In contrast, the amphipods represent the 'Swiss Army knife' of the crustaceans. There are plenty of examples where different legs on the same animal can be used in feeding, breathing, grasping, biting, jumping, swimming, brushing and clasping the opposite sex. While a centimetre is a common length, a few males of select species can be double that. As with the isopods, their eyes have no need to be compound as the animals remain by day under stones, rocks and seaweeds. The common name for amphipods is sandhoppers, which is rather ill-fitting since most are not found on sand and do not hop. The typical sandhoppers occur in many thousands on the strandline, and are therefore discussed in Chapter 12.

Gammarus locusta is the archetypal amphipod, with a long flattened body and the tail curved underneath. It belongs to the family Gammaridae, the species of which have a distinct branch on the first antennae, quite visible when submerged in water. A male *G. locusta* can grow to 20mm, almost twice the length of the female, which he will often carry, grasped under his body, for a couple of days before mating. When she lays her eggs they are retained in a brood pouch until they hatch. This species and other amphipods are important detritivores, clearing up the fine detritus and debris trapped by the shore as the tide retreats. In turn, they are a valuable food source for crabs and fish.

The less visible caprellid amphipods are rather charming. Known as skeleton shrimps, these are very slender creatures and have a stance like a praying mantis, remaining quite still for long periods attached just by the claws on the hind legs. The common Ghost Shrimp

Rocky Shores

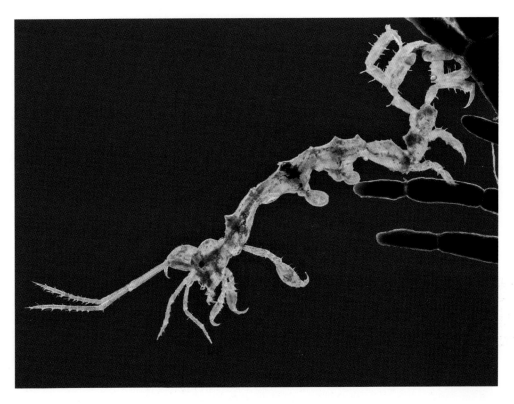

Caprellids are like ghostly skeletons clinging onto seaweed as they feed on suspended matter in the water.

Caprella linearis can be almost transparent, or it may take on the colour of its surroundings: usually a small red seaweed or a colony of hydroids and bryozoans. Like so many small marine organisms, caprellids are best found by placing a small amount of the seaweed in a dish of seawater and waiting for life to emerge. *Caprella* have particularly large and noticeable pincers for collecting food.

An even stranger-looking amphipod is *Hyperia galba*. Sizes are comparable with *G. locusta* but here the female is the larger of the two sexes at 20mm. They are rather rounded and have huge eyes, living out at sea away from the shore during the winter. If you want to see one, look for a stranded jellyfish, especially the Barrel Jellyfish *Rhizostoma pulmo*. The amphipod infests the host for much of the year, residing inside the cavity beneath the bell of the jellyfish where it eats the host, debris caught up inside the bell and possibly particles of the host's food. The female produces large numbers of young that tend to stay within the protection of the host until they are sufficiently developed to swim away and find their own host, which can take several months.

Decapods

Amphipods are often informally referred to as shrimps, but the true shrimp forms are restricted to the large, diverse and spectacular order, the Decapoda.

The books I (JC) read as a child are mostly a distant memory, but one example stands out from when I was about five years old – Enid Blyton's *Noddy Goes to the Seaside*. In my mind's eye I can clearly see the illustration of Noddy being nipped by a crab as he lifts his foot out of the sea. I seem to remember that the eyes on the crab were red, a characteristic of the beautiful blue-jointed Velvet Swimming Crab *Necora puber*. If there is one crab liable to turn and fight while the others scuttle away or curl up in a defensive ball, it is that one. I once witnessed an unsuspecting student poke a young, stationary Velvet Swimming Crab with a pencil; rather than withdraw, the crab grabbed the pencil and gradually squeezed until it was almost completely snapped in two. That is the power behind the claws. The chelae, to give them their proper name, are actually the first pair of the crab's ten legs; the remaining eight are the walking legs. Being all of similar length, they would cause their owner to trip over if it weren't for the crab's classic sideways locomotion. Not all decapods walk, however, and a simple means of dividing the order is into the walkers (Reptantia – crabs and lobsters) and the swimmers (Natantia – prawns and shrimps).

A gammarid shrimp among coral weed, with a tubeworm feeding at the top of the image (note the 'snow-storm' of detritus in the water).

The Velvet Swimming Crab is most likely to be encountered under rocks in the lower shore. It can travel quickly through rock pools using its last pair of legs, which are modified into flattened paddles for swimming. If you dare, stroke the top of the carapace and you'll find a surprising fine down of velvet. This is lacking in other species. The nearest in hirsuteness is the small (25mm) Hairy Crab *Pilumnus hirtellus*, which has a coating of much longer hair and is more likely to be found inside the holdfast of large kelp plants.

While the Velvet Swimming Crab is the most colourful representative of the decapods, the most menacing is probably the Marbled or Montagu's Crab *Xantho hydrophilus*. Despite its appearance, it is actually quite slow and easy to pick up by grasping the sides of the carapace. It is a lower-shore species and more likely to be found in the south or west of Britain and Ireland.

The related Edible Crab *Cancer pagurus* has a propensity, when disturbed, to fold its legs and chelae neatly into a defensive ball. Distinguished by the piecrust edge to the carapace, it is only found on the shore in the younger phase of its life, when it is up to 100mm wide. As it feeds and develops on the shore, so it moves slowly into deeper water until away from the tidal influence. Several long journeys have been recorded, including one individual that started by being tagged in Whitby and was subsequently caught off northern Scotland, more

A Marbled Crab, with a number of tubeworms growing on its exoskeleton.

Arthropods: animals with jointed limbs

An Edible Crab, showing the characteristic 'pie-crust' edge to the carapace.

than 400km away. The Edible Crab reaches maturity in around five years, and continues to grow, eventually becoming the sort of giant that appears in fishmongers. In 2011, 12,000 tonnes were landed in England and Wales, making this species the most valuable crustacean in the UK.

Few animal species on the seashore can match the resilience, adaptability and physiological tolerances of the Common Shore Crab. No wonder that it is the most abundant and widespread crab in the British Isles. Unlike the other decapods mentioned so far, it is not restricted to the lower shore. Some smaller individuals can be found as high as the upper shore; typically numbers increase in the middle shore but decrease in the lower shore because of the high levels of interspecific competition that exist there, where conditions are most stable. The main stresses higher up are of course temperature variation and desiccation. In addition, any standing seawater at low tide is likely to be hypersaline because of evaporation, or diluted by rain and freshwater streams. While all crustaceans possess a pair of antennal glands located near the base of the antennae, in the Common Shore Crab these are able to provide good osmoregulation. Being able to regulate the ionic concentration of the haemolymph enables the crab to exist in estuaries with salinities as low as four parts of sodium chloride per thousand of water (4‰; normal seawater is 34–35‰). Behavioural adaptations can also help. We have seen these crabs find the deepest part of a brackish pool with a freshwater stream flowing

A Common Shore Crab in a rock pool. Note the short antennae, typical of true crabs.

through; seawater with its higher density sinks to this lower area and the freshwater floats over the top.

During the 1960s the biologist R. Binns at Newcastle University carried out much of the research on the Common Shore Crab's antennal glands. He showed that despite the general consensus that they were like kidneys, the urine flowing from them actually had a *low* content of nitrogenous waste, barely 5–10 per cent of all of such waste produced by the crab. In all probability most crustaceans produce ammonia that diffuses out of the body without the need for specialist organs. Interestingly, Common Shore Crab colour seems to vary, and this has been ascribed to either salinity or age, or both. The most common colour is green, with a tendency to get darker as individuals age. Pale varieties occur on the lower shore and in deeper water.

Crabs are not obviously segmented unless viewed from below. With the tough carapace covering the head and thorax, the body is slightly flattened and the substantially reduced abdomen is folded neatly up underneath. The abdomen of the female is distinctly broader than the male's and has seven visible segments versus his five. Males will hunt out females in the summer prior to mating; identifying a mate that is about to moult is crucial, as only when the exoskeleton is soft after moulting will she be able to copulate. Once they find each other they stay together until the female moults, the male helping the female to extract herself from the old skeleton. The female Edible Crab is known to release a

Arthropods: animals with jointed limbs

A Common Shore Crab female carrying eggs under the abdomen.

secretion that hardens on contact with sperm to form a plug, preventing the loss of sperm in the seawater and stopping other males trying to copulate. This may well occur in other species. Once the eggs have been laid, the female retains them attached under her abdomen, referred to as being *in beri*, and she can be loaded with around 150,000 red eggs that hatch after some weeks into the planktonic zoaea larval stage.

Next time you pick up a shore crab by the sides of the carapace, be brave and stare it straight in the face. In no time, you will see evidence of another adaptation: it will begin blowing bubbles from its mouth as it breathes air. Now try holding the crab so that the back of the carapace is dipped into water, but keep looking at its face. Notice how the bubbles stop and water dribbles out instead. You have just demonstrated how crabs breathe, using balers that draw water (or air, when emersed) through a gap below the back of the carapace and through the gills that fill the cavity under the carapace, to exit through the mouth. We demonstrate this in a dish of seawater by dropping a small amount of blue food colouring at the back of the crab and watching the coloured water stream out of its mouth.

Many of the crustaceans discussed in this chapter are capable of regenerating limbs by a process called autotomy, and for species like the Common Shore Crab, which often lives under small, mobile rocks, this is especially useful. If a leg becomes caught under a rock, or grasped by a predator, the crab can snap it off at a predetermined break point.

A large gathering of the spider crab *Maia squinado*.

Inside, near the base of the leg, is a double membrane through which nerves, blood and a muscle penetrate. On contracting the muscle, the leg breaks free with a membrane remaining on the body segment to prevent blood loss. At the next moult, a new limb will be grown.

Rarely do crabs make headline news, but in June 2003 the BBC reported the 'biggest influx in living memory' of the spider crab *Maia squinado* into Cardigan Bay. It was unclear whether climate change, population explosion or a change in ocean currents was the cause of thousands of south Wales's spider crabs moving into the bay for the summer. In May 2016 spider crabs hit the headlines again, following a mass movement into St Bride's Bay, Pembrokeshire. While some newspapers reported an 'alien invasion', others recognised a movement of crabs inshore to moult and mate. Many arrive on land at low tide and are visible in sandy patches between rocks or caught in deep rock pools amongst kelp. A writhing mass of large spider crabs is quite a sight, but often results in a number of dead bodies cast up onto the shore. Increasingly, people are coming to film the annual migration of the crabs for the sheer numbers involved and the sensational size the animal can attain. *Maia squinado* is the largest arthropod in the British Isles: a typical female might boast a carapace length of 128mm.

Maia may be the most spectacular of the spider crabs, but several much smaller species are encountered in lower-shore pools. They have

Arthropods: animals with jointed limbs

A spider crab, *Inachus dorsettensis*, covered in pieces of red seaweed that it has cut and attached to its exoskeleton.

a triangular-shaped carapace: *Hyas araneus* is common in the south and west while the delicate-looking *Inachus dorsettensis* is more northerly. Both are closer to 30–40mm across and camouflage themselves by sticking pieces of seaweed all over their carapace and legs. This has led them to being commonly known as decorator crabs. *Inachus* is omnivorous but *Hyas* is a seaweed consumer. Both are slow-moving and hide amongst the fringe weeds in pools.

A common sight under rocks in the Serrated Wrack zone is a creature that appears to be a crab, the so-called Broad-clawed Porcelain Crab *Porcellana platycheles*. There is also a Long-clawed Porcelain Crab *Pisidia longicornis*, but it is less abundant. At times, the former is so common that dozens can be seen scrabbling about on the undersides of rocks when they are lifted for inspection. The nature of their movement may give away the fact that they are more lobster than crab: they flip the abdomen – which is located under the body – rapidly to move away from possible danger when in water. In addition, they have noticeable long, whip-like antennae (true crabs have short antennae). Their chelae are covered in a dense coating of hair, which they apparently use to brush debris from the rock. This, along with other detritus, is filter-fed into the mouth – quite different from the carnivorous and scavenging nature of the true crabs. A porcelain crab might also surprise you when you first pick it up by immediately gripping on tight with its legs. The

The Broad-clawed Porcelain Crab has a flattened body, enabling it to live in large numbers under stones.

sharp claw at the end of each leg provides exceptional grip on rocks (or hands), so they are difficult to remove, probably an adaptation for hanging on during wave action. They are also very flat, so they can easily hide under rocks and in crevices.

Being able to flip their abdomen rapidly under the body is a good escape mechanism for crustaceans. The Lobster *Homarus gammarus* will do this but is not often seen on the shore, and if the occasional individual does appear it will be a young juvenile. More likely to be seen are the squat lobsters, such as the greenish-coloured 40mm-long Black Squat Lobster *Galathea squamifera*, found under rocks on the lower shore. The related Spiny Squat Lobster *G. strigosa* is a much flashier red with sharp spines over the body. Be careful picking these ones up, as they can be aggressive when disturbed and grow to 120mm.

Strangely, like the porcelain crabs, hermit crabs are not classified as true crabs. They are familiar to most people because of their habit of covering the body in a gastropod shell, starting with a small periwinkle shell and exchanging this regularly for successively larger ones. The crab is very particular in its choice of home, checking a potential new shell thoroughly for weight and size before making the final decision to relocate. Hermit crabs have a tough head and thorax, while the very soft abdomen is asymmetrical, with claspers at the end to hold

Arthropods: animals with jointed limbs

The Spiny Squat Lobster lives in the sublittoral zone but occasionally appears on the shore.

the shell. Over time their shell is colonised by hydroids, sponges and other filter-feeding animals, as hermit crabs are messy feeders and particulate matter regularly drifts away. The Common Hermit Crab *Pagurus bernhardus* routinely has a specific hydroid, *Hydractinia echinata*, and occasionally an anemone, *Calliactis parasitica*, on its shell. Despite the name of the anemone, it is not a parasite but an associate. Each of the pair could survive alone but develops better with the partner: food for the anemone and some degree of protection for the crab. This relationship is an example of mutualism, whereby each member of the association benefits, as we saw with the lichens in Chapter 4. Inside the shell on the back of the hermit crab may be found the polychaete worm *Nereis fucata*.

Rocky Shores

A hermit crab with the hydroid *Hydractinia* on its shell.

Hermit crabs can become abundant in a suitable middle-shore pool, with dozens scurrying around on the rock-pool bottom. Scavenging on dead organic material left in a pool by the receding tide, they will have to compete with the prawns, the best example of swimming decapods. Species such as the Common Prawn *Palaemon serratus* are quite territorial and will defend their patch of the pool.

European Glass Prawn *Palaemon elegans*, a common rock-pool species, with eggs beneath the abdomen.

A prawn's body is more flexible than a crab's, because it has a thinner exoskeleton, and this allows a rapid response, a flick of the abdomen resulting in a quick retreat. Swimming across open water requires good senses. Four long antennae extend outwards to the front, sides and below, reacting to touch and taste. The bases of the antennae are flattened plates that act as stabilisers in the water. As in all decapods, the eyes are compound and very well developed on moveable stalks, providing an excellent 180-degree view. The lenses are square and are very sensitive to movement. Also generally visible near the eye are small star-like chromatophores (pigment cells), which can make changes to the body colour and also reflect colour. Found in a number of crustaceans, they are more obvious in the prawns because of their transparent skeletons.

The terms prawn and shrimp are often used interchangeably. However, biologists generally recognise that prawn is the name given to those species that have an extension of the carapace like a serrated knife, called the rostrum, lying between the eyes and protruding some way forward from the head. The mysids, mentioned earlier in this chapter, and the decapod species such as *Crangon* that bury themselves in sand, lack the rostrum and are shrimps.

* * *

Many of the arthropods, and indeed many other animals that live on the seashore, have restricted powers of locomotion. But they have a secret weapon that increases their dispersal ability – a floating (planktonic) larval stage. It is to this phase of the rocky-shore story that we turn our attention next.

Despite its jointed limbs, in comparison with the thinner prawn's body the hard carapace and pincers make this Edible Crab slow to respond to predators.

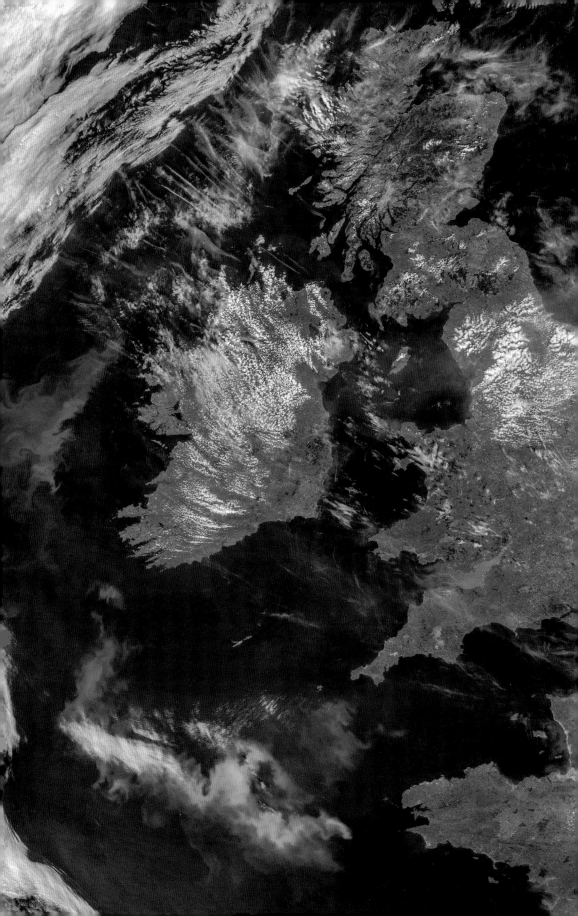

Plankton: drifters of the shore

chapter ten

'I can give you no more idea of what the species are like further than that the magnified figures resemble the objects at the far end of a kaleidoscope'

Botanist and explorer Sir Joseph Hooker, in a letter to Captain Sir James Ross (1844)

May was a lovely month in Pembrokeshire as we were writing this book. When June arrived so did the wind, coming strong across the Atlantic. The blow lasted four days and then all was calm again. It seemed that every rock pool in St Bride's Bay had a small, blue boat sailing around on its surface; they were in fact the colonial hydroids of the beautiful By-the-wind Sailor *Velella velella*, which is related to the Portuguese Man O' War. Reaching 80mm across, the deep blue oval discs consist of a horny skeleton enclosing a float. Hanging below is a series of hydroid zooids or polyps; the central ones are for feeding, and reproductive polyps surround them. A distinct diagonally arranged horny sail sits on the top of the oval disk. The specimens we observed had been blown in from the warm ocean waters of the tropics, where they live. Mass strandings can be common in late summer and autumn on the Atlantic coast. While the sail is useful to pick up the wind and aid dispersal of the colony, individuals do not have any control over their movement or direction. These drifters at the mercy of the tide, currents and wind are examples of a disparate group of organisms, termed plankton, and are the antithesis of the active swimmers, such as shark or cod, which are sometimes referred to as nekton.

OPPOSITE PAGE: In this satellite image, a phytoplankton bloom colours the ocean blue and green.

BELOW: A By-the-wind Sailor from the open ocean, stranded on the surface of a rock pool.

Jellies

In Chapter 6 we met the colonial hydroids, tiny polyps attached to seaweeds or rock. To reproduce, hydroid polyps shed a microscopic form called a medusa, complete with gonads, and this short-lived sexual phase releases gametes into the water, where the fertilised egg hatches into a larva called a planula. Jellyfish are like the medusa phase of hydroids. As well as being similar in appearance, they contain gonads and are the reproductive phase. They also produce planula larvae. But in the case of jellyfish the larva does not become anchored as a polyp. The free-floating medusa – typically much larger than the hydroid medusa – is the final adult stage.

Jellyfish may fit the 'drifter' definition, but by planktonic standards they are large. They typically come to our attention when they are abandoned by the tide, often stranded in rock pools. The limestone shores of Galway Bay can be a fruitful place to search for large jellyfish funnelled in by the topography in summer months. My (JC) first Compass Jellyfish *Chrysaora hysoscella* was a huge individual swimming as best it could in a large pool on the edge of the Burren. It was the first blue-sky day after several weeks of constant bad weather and the shore was dotted with both these and the Common or Moon Jellyfish *Aurelia aurita*. Like other cnidarian species, they have characteristic sting cells on the tentacles for catching prey. The Common Jellyfish

A Compass Jellyfish trapped in a tidal pool.

have reduced tentacles, and although they do catch larval fish they usually use streams of mucus over the body to collect microscopic plankton, using small appendages to move particles to the mouth.

Smaller than jellyfish are the sea-gooseberries, some of the most beautiful creatures one can encounter trapped in rock pools at low water. Their phylum, Ctenophora, means 'comb-bearer' in Greek; they are also known as comb jellies, as they appear to have eight 'combs' running down a gelatinous body. These are rows of cilia, for movement, which create stunning rainbow-coloured interference patterns in the light as they move. Otherwise, ctenophores are colourless, almost transparent and very difficult to see. They swim mouth first, with a voracious habit of consuming everything in front of them: usually crustaceans and other creatures living in the plankton. Although superficially they look like a jellyfish (and used to be classified with them), no one really knows what they are related to. Recent genome studies suggest that they could be linked to sponges.

A drop in the ocean

Nothing beats an exhilarating swim in the sea to reconnect with nature, but how many bathers ever stop to consider what they will inevitably ingest during the course of their dip? Plenty of minerals, certainly, microscopic plankton, and probably a mass of sperm and eggs produced by various invertebrates and seaweeds! In all likelihood, one drop will not contain a very diverse plankton sample, which is why when we look for plankton we need a proper net or nets to concentrate the sample. Different nets have different mesh sizes. Holes of 150 microns (the average kitchen flour sieve has a mesh ten times that size) are relatively big and retain only the larger animals (zooplankton) along with a few of the large plants (phytoplankton) or floating algae. A 60-micron mesh size will trap the bulk of the zooplankton and a very good proportion of phytoplankton. A hand net deployed from a jetty will provide small samples, but the best technique is to tow the net behind a boat. This needs to be done slowly, especially with finer mesh sizes, due to the 'bow wave' that builds up in front of the net pushing the water and plankton away from the net opening. Plankton nets have a collecting tube or bottle at the apex, which ideally can be unscrewed and the contents poured into a sample bottle so that more samples can be taken. Samples need to be kept cool and examined as quickly as possible. A large body of water has a stable

 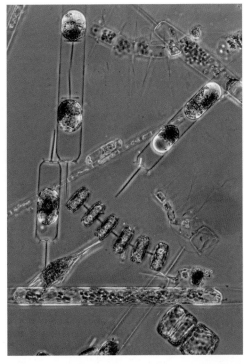

ABOVE: A phytoplankton net being hauled in after a tow. The bottle at the end holds the sample and can be unscrewed to retrieve it.

ABOVE RIGHT: Phytoplankton sample.

temperature, but that of a small sample will fluctuate rapidly, killing the life inside. Lack of oxygen is another problem as the multitude of organisms inside the jar respire, and it is likely that a sample will also contain plenty of carnivorous animals, which will quickly reduce the assemblage to a few satiated individuals.

A stereo microscope, which gives a relatively low-power view, is a good starting point for examining the sample in a shallow dish, and then smaller drops taken from this can be transferred to a glass slide and illuminated from beneath so they can be viewed in more detail under a compound microscope. Samples taken at midday in April or May could take days to analyse as there will be so much material to examine; keeping them refrigerated helps to maintain freshness for a day or so, but after that a preservative will be required.

A sample of plankton soup will contain plenty of detritus, most of which will be difficult to identify. Minute pieces of seaweed may be covered in shimmering black dots: these are bacteria feeding. Large amounts of shed crustacean exoskeletons, especially barnacle cirri (the legs they use to collect food), are also likely to be present. Otherwise, the division is fairly clear between phytoplankton and zooplankton – although we will come to some of the oddities in due course.

Crinoline to silk – a history of plankton research

The pioneer microscopist Antonie van Leeuwenhoek is believed to have been the first person to see and describe a diatom. In 1676 he wrote a letter to the Royal Society of London to tell them that while looking at seawater on the Dutch coast at Scheveningen, he found some very small creatures. Although there are no drawings to accompany the descriptions, he did explain in some detail what he saw under his microscope. Almost two centuries later, Joseph Hooker, the naturalist friend of Darwin, accompanied the *Erebus* on the Antarctic expedition as ship's naturalist and dabbled in plankton sampling. However, much of the work around this time was carried out by German scientists. Regular plankton sampling and studies were initiated in the small Heligoland archipelago in the North Sea in 1845 by Johannes Müller (and work continues there today). Müller did not call the objects of his study 'plankton' – that term would appear much later – but *Auftrieb*, which translates as 'lift' or 'buoyancy' and presumably refers to the floating ability of his subjects. Young German naturalist Ernst Haeckel visited Heligoland and later recalled how in 1854 Müller was so excited by plankton he said to him, 'as soon as you have entered into this pelagic wonderland you will see that you cannot leave it.' Haeckel went on to describe thousands of new species.

Meanwhile, also in the 19th century, the marine biologist G. C. Wallich dropped some nets made from crinoline overboard from his ship and collected plankton samples while crossing the Atlantic. Later, in 1877, he described the diatom *Coccosphaera pelagica* (in 1930 the genus was changed to *Coccolithus*).

Professor Victor Hensen in Germany is probably the most important researcher when it comes to putting diatoms and plankton on the map. It was he who first coined the term 'plankton' (in 1887), although he meant this to refer to anything that floated around in the sea, dead or alive. In a very forward-thinking manifesto, he suggested that the fisheries industry was totally dependent on understanding fish larvae and their reliance on the plankton for food, and that the productivity of plankton would determine the productivity of fish stocks. To investigate this in the Atlantic he set up the Plankton Expedition (also referred to as the German Expedition). In 1889 the

Diatoms of the genus *Coscinodiscus* are some of the largest found in the phytoplankton.

steamer *National* left the Prussian city of Kiel with Hensen and fellow scientists for a three-month figure-of-eight voyage of 25,000km, taking in the North Sea, Greenland, Bermuda, the Amazon region and the English Channel. Seasickness and equipment failure caused numerous problems for them, but this was the first detailed analysis of the ocean's drifting life. Hensen developed and applied scientific methodology to the sampling so that it could be repeated. Later, Ernst Haeckel used the term plankton in his writing in 1890, but never referred to dead matter. From then on, only living organisms were considered as plankton. By this time silk had replaced crinoline as the material for making plankton nets, as it was fine enough to pick up microscopic diatoms and dinoflagellates.

During the 19th century most research on plankton was carried out in continental Europe. The British *Challenger* marine expedition of 1872–1876 studied the oceans of the world but, surprisingly, carried out very little research on plankton. All this was to change in 1925, when Alister Hardy (later Sir Alister) joined the RRS *Discovery* on a two-year exploration of the Antarctic. Although later in life he was to become a world-renowned expert on whales, at this time it was plankton that inspired him, and during his time on *Discovery* he designed the Continuous Plankton Recorder or CPR. This device uses a moving band of silk on which the plankton is trapped and stored in preservative for later examination, a mechanism still used today. In 1956 the first of his two New Naturalist books on *The Open Sea* was published, covering 'the world of plankton'. Nothing like this had been printed before, and even the editors were unsure how to describe it. On the inside of the jacket they wrote that they believed it to be 'the greatest general work on the subject ever written'. It remains one of the most readable, fascinating and inspirational works on the subject. Hardy's pioneering work is continued today by the Sir Alister Hardy Foundation for Ocean Science, based in Plymouth.

Diatoms and dinoflagellates

Diatoms are a division of the green algae, the Bacillariophyta, and estimates of diatom species diversity run into hundreds of thousands. Typically they are single, solitary cells, but some, such as *Thalassiosira*, form into chains. The exquisite nature of a diatom comes from the outer wall, known as the frustule, which is composed of silica, like clear glass, enabling light to pass through for photosynthesis. The

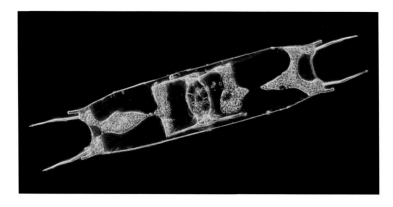

An *Odontella* diatom dividing into two.

frustule is made of two halves or valves, and in the round diatoms it is easiest to imagine them being like the two halves of a petri dish with one overlapping the other.

Shape is a simple way to classify the diatoms into two main groups, but even here diversity creates overlap and difficulty. A common genus, *Coscinodiscus*, has large individuals up to 500 microns in diameter and is a good example of the first type, the circular or centric diatoms. The second group consists of the pennate forms, which are bilaterally symmetrical with a line or raphe down the middle. The long, narrow cells of the pennate *Bacillaria paxillifer*, for instance, link along their length to create a mobile colony. The cells slide along each other, constantly changing the shape of the aggregate, and it has received common names such as Sliding Diatom and the Carpenter's Rule as a result. Highly magnified diatoms display siliceous structures in the frustule called fibulae, a series of lines and canals. Microscopists today use these diatom fibulae as a way of measuring the resolving power of their lenses.

As a rule, Victorian scientists wanted to bring order to the natural world. For most, this involved naming and classifying; but for some, beauty and form were what mattered. From the moment the intricacy of diatoms was recognised, a craze developed for turning species variety into an art form. Johann Diedrich Möller (1844–1907) is perhaps the most famous producer of exhibition art, and he collected diatoms to arrange and mount on microscope slides. Many emulated him, and the fashion reached a peak at the beginning of the 20th century. While the techniques have largely disappeared, the slides are still in circulation today. Only one microscopist, the Englishman Klaus Kemp, continues to produce diatom art in the 21st century, artfully placing the kaleidoscope of diatom cells to create intricate,

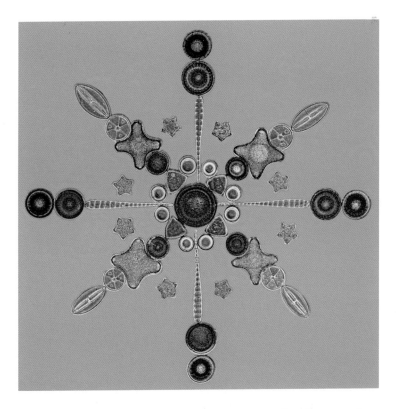

An exhibition rosette of 49 marine diatoms, arranged and displayed by Klaus Kemp.

symmetrical patterns. Some of his creations look like flowers; others may resemble more complex objects, such as a bicycle or an American eagle (see www.diatoms.co.uk).

A sample of phytoplankton can yield some strange creatures that defy attempts to classify them as 'plants' or 'animals': these are the dinoflagellates. In this group, a pair of flagella arise from two grooves in the surface, enabling movement and creating a feeding current in some species. This places them in the (unsatisfactory) kingdom of protists, as they can both photosynthesise and collect organic food. Similar in size to the diatoms, they can be split into thecate and naked forms. The former have the cell covered in a series of cellulose plates – the theca – while the latter, without these plates, can be more variable in shape. The majority of dinoflagellates are marine, and some, the zooxanthellae, are endosymbiotic, meaning that they live inside other living organisms. The common Snakelocks Anemone that we met in Chapter 3 has both a grey and a green form, often living side by side in a rock pool; the green colour is due to the dinoflagellate living inside that contributes the products of photosynthesis to the anemone, which in turn provides shelter and carbon dioxide.

Blooming productivity

Phytoplankton species reproduce asexually by binary fission, with each cell dividing into two. In diatoms, one valve goes to each daughter cell, plus half of the contents. A new valve then grows to replace the missing one, and as this fits inside the other a smaller diatom will result from the smaller half. Over time a very mixed array of sizes will be present in the population. This is corrected eventually if sexual reproduction occurs, when a spore is produced that ultimately develops into a new diatom. Binary fission can occur up to four times a day in a diatom, as long as all the necessary nutrients such as silica and phosphate are available. The main sources of minerals lie at the bottom of the sea (the benthos), because when organisms die that is where they descend. To lift these nutrients into the sunlit upper areas of the sea, some disturbance such as a storm is needed. In the shallow seas around the British Isles, wind-induced mixing of the water, aided by strong tidal currents and uneven sea floors, will stir the benthos and bring the nutrients up to where they can be used by the plants. Phytoplankton density builds in spring and autumn when strong winds and gales are at their peak, with the spring bloom – when sunlight is most abundant and sea temperature is rising – being the largest and longest. Plankton is at its most dense in water less than 200m deep.

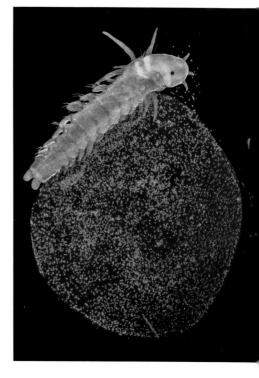

A gelatinous colony of the diatom *Phaeocystis globosa* with a polychaete larva of a paddle worm, possibly *Phyllodoce*. A bloom of these colonies can cause significant problems, creating huge amounts of gelatinous matter stranded on beaches.

Blooms of phytoplankton can be so large that they are visible from space (see p. 268). The huge productivity of biomass on this scale is unequalled on our planet, and given our concern for high levels of carbon in the atmosphere, it is worth appreciating that phytoplankton has an important role to play as a carbon sink, albeit a temporary one as carbon makes its way through the marine food web. The phrase 'standing crop' refers to the total amount of living matter in a specific area at a given time; compare coastal plankton with a wheat field ready for harvest, and you could be forgiven for thinking that the field system is the more productive of the two. By comparing net productivity over an annual period, however, it becomes apparent that parts of the ocean can be considerably more productive than the land. A simple diatom duplicating itself four times a day is an almost inexhaustible food resource.

Phytoplankton blooms are not necessarily all good news. In May 2016, a fairly rare event was observed in Milford Haven when the diatom *Phaeocystis globosa* (see p. 277) bloomed. As we lifted the plankton net out of the sea at the back of the boat, very little water poured through the net wall; the mesh and bag of the net were full of a slimy deposit. When *Phaeocystis* blooms it enters a colonial phase where cells cluster around a secretion of mucus. Individual globules of mucus can be centimetres across, and when hundreds are collected in the net it soon becomes a bag of slime. If large amounts wash up on beaches the diatom becomes a serious nuisance, and at sea the death of the bloom creates areas that are low in oxygen.

Dinoflagellates can make themselves very obvious on the occasion of a bloom. Many are bioluminescent, so when they are disturbed at night by waves or a boat an eerie glow appears in the water. At times, when nutrient availability, light and temperature coincide to create the perfect cocktail, a rapid growth and bloom of dinoflagellates can occur that generates a red-coloured water that is highly toxic: a red tide. Toxins produced by dinoflagellates can be concentrated in the tissues of shellfish and other invertebrates, and this has led to a number of cultural taboos among indigenous peoples around the world; on the northwest coast of America native tribes would ban the collection of any shore creatures for food if they saw lights in the water.

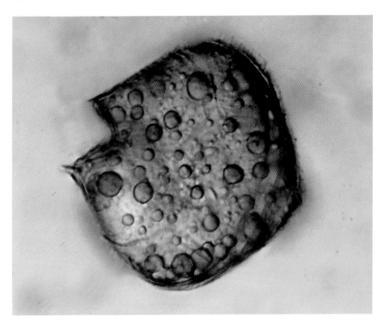

Protoperidinium, a dinoflagellate.

Zooplankton and metamorphosis

In contrast to the situation for phytoplankton, the microscopic animal life that drifted in the seas around the British Isles was well documented even at the start of the 19th century, owing in part to some zooplankton being just visible to the naked eye. That is the good news. Unfortunately, from the perspective of modern science, much of it was misidentified. Today we know that a crab produces a larva that swims off to spend the first part of its life drifting through the plankton, but 200 years ago these amazing-looking creatures that appeared looking nothing like the adult were thought to be new species. Otto Müller described the genus *Nauplius* in 1785; Louis Bosc discovered what he thought was a new genus, *Zoaea*, in 1802; and William Elford (at the time the leading authority on crustaceans) founded the genus *Megalopa* in 1813. Subsequent research has demonstrated that the nauplius is actually the young larval stage of a barnacle or copepod (as we saw in Chapter 9), while a zoaea is the first stage of a crab, which goes on to form a megalopa prior to being an adult.

From the 1840s onwards, naturalists began to realise that they were witnessing metamorphosis in marine creatures. Darwin was one of the first, when he realised that barnacles came from a nauplius. Philip Henry Gosse concurred with Darwin on the development of barnacle larvae in his book *Tenby* (published in 1856), in which he described watching

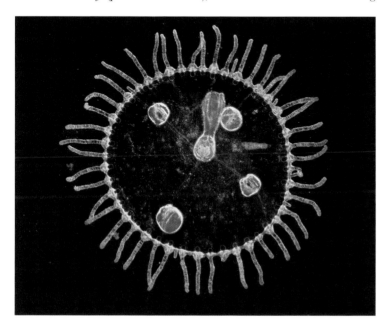

A medusa of the hydroid *Obelia*. Medusae are the sexual and dispersal phase that detach from the polyp growth stage found on the rocky shore.

several weird creatures metamorphose into squat lobsters. These are some of the first references to larval stages living in the plankton.

Metamorphosis is a classic feature of the development of other familiar arthropods such as the insects. Some groups (butterflies, for example) undergo complete metamorphosis (holometabolism), in which they go through larval (caterpillar) and pupal (chrysalis) phases before reaching adulthood, while others (such as grasshoppers) have an incomplete metamorphosis (hemimetabolism) in which a series of individuals produces stages or instars that look increasingly like the adult. Metamorphosis is also a key feature of many marine invertebrates – and it is no wonder that their larval stages were so often misidentified, considering how dramatically their body forms can change through the process.

For sessile invertebrates living on the shore, dispersal of their young presents a major problem. A period spent living in the plankton is the ideal solution, and this is why so many rocky-shore creatures have planktonic young, but there are downsides to this strategy, since the young may get lost out at sea and die before reaching a suitable environment, or find themselves prey to the huge numbers of predators that live out there.

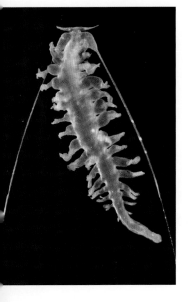

A beautifully transparent *Tomopteris* species, a polychaete that lives permanently in the zooplankton.

Some members of the plankton spend their whole lives within the drifter community and are referred to as holoplankton (see p. 286); obviously none of these will be typical rocky-shore residents. Members of the plankton that only spend part of their lives drifting before dropping to the substrate, such as onto a rocky shore, are called meroplankton. Others, like fish larvae, become active swimmers once out of the larval stage and join the nekton. Animals of the shore will often synchronise their larval release to coincide with algal blooms, as is the case in some barnacle species in March and April, while others may shun peak periods to avoid intense competition for resources. Limpets, for example, spawn in late October.

From an array of names for the different stages of different animal groups, we will mention just a dozen or so here. The simplest larval form is the planula larva: a flattened, ovoid clump of cells covered in beating cilia that settles on the shore to form a cnidarian such as a sea anemone (although identifying the species to which a planula belongs is next to impossible). Marine flatworms have a number of larvae, the group known as Müller's larvae being the commonest around our shores. Slightly ovoid, they have several lobes covered with cilia and all sited around a small opening, the mouth, for consuming smaller larvae and diatoms.

Some planktonic larval forms

Larva	Phylum/class and group
Auricularia	Holothuroidea (sea-cucumbers)
Bipinnaria	Echinodermata
Brachiolaria	Asteroidea (starfish)
Cyphonautes	Bryozoa
Cyprid	Crustacea (barnacles)
Megalopa	Crustacea (decapods)
Müller's larva	Platyhelminthes
Mysis	Crustacea
Nauplius	Crustacea (barnacles and copepods)
Nectochaete	Polychaeta (some groups)
Ophiopluteus	Ophiuroidea (brittle-stars)
Pilidium	Nemertea (ribbon worms)
Planula	Cnidaria
Pluteus or Echinopluteus	Echinoidea (sea urchins)
Trochophore	Polychaeta, Mollusca
Veliger	Mollusca
Zoaea	Crustacea (decapods)

Another group, the trochophores, are the first stage in the development of marine worms or polychaetes. They are similar to the Müller's larva but have a mouth and an anus. The keelworm *Spirobranchus triqueter* has an especially fine example of the trochophore with a band of cilia around the middle so that it spins like a top. From this early stage the trochophore develops into another larval phase with bristles. In contrast, molluscs have a veliger larva. Each stage in its development comes with increasing complexity, starting off as a trochophore sphere that quickly secretes a simple protective shell, followed by a wide velum (a bit like wings) covered in cilia to hold the increased weight. The metamorphosis of a fully grown veliger is quite a spectacle to behold as it loses the velum and settles to the sea bottom and changes into a young adult marine snail or mussel. The Edible Periwinkle veliger can commonly be found in the spring plankton.

There is one larva that when first viewed under a microscope conjures up images of a gymnast star-jumping. This sea urchin larva is one of many pluteus types found among the echinoderms. It generally starts off as a bipinnaria larva, like an extended trochophore

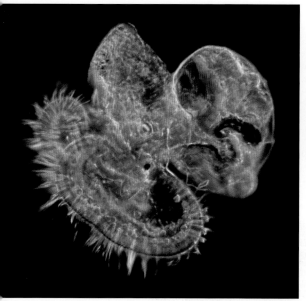 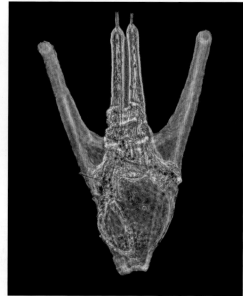

LEFT: The veliger larva of a periwinkle.

RIGHT: The pluteus larva of a sea urchin.

but bilaterally symmetrical. The challenge all planktonic larvae face is remaining buoyant and not dropping like a stone to the bottom. Extensions and lobes covered with cilia help, but as larvae age and grow they must add more projections, and hence surface area, to keep them in the plankton-rich surface layers. Eventually the only option for the bipinnaria is to develop into a form that has many long extensions, as in the pluteus, to provide buoyancy.

The development of body extensions is carried to an extreme in the case of the crab zoaea larva. Crustaceans and their larvae do not possess any cilia that they might beat to keep them stable in the water column, so zoaeas have an exoskeleton with jointed limbs for locomotion and a segmented body. This requires the zoaea to moult four or five times as it grows. Compound stalked eyes are typical, as is a carapace over the head and thorax, with two substantial spines extending in different directions to provide stability so that the larva does not roll over as it swims. In some decapods, such as the porcelain crabs, the spines are very long both to the front and to the rear of the body. On the final moult, the zoaea changes shape to become a megalopa larva. This new crab-like flattened body has a broad carapace with the abdomen projecting to the rear. It swims with its legs but quite soon settles on the sea floor, where it completes metamorphosis by moulting to bring the abdomen under the carapace and become a young adult.

Plankton: drifters of the shore

LEFT: The zoaea larva of a crab.

RIGHT: A late-stage nauplius larva.

Shrimps and prawns have a different juvenile stage from the crabs, in which the megalopa is replaced by a mysis larva, so-called because it is very similar to the primitive mysid crustaceans and may even be an evolutionary 'throwback'. Lobsters have a mysis larva from the start, as the zoaea larva occurs inside the egg.

As previously mentioned, barnacles have a very different larval form that is very common in the plankton: the nauplius. Oval to triangular in shape, it has a single eye at the front of the head, which is at the broad end of the unsegmented body. Either side of the head are horns that link to a pair of glands. Swimming is achieved with three pairs of hairy appendages, but feeding does not start until the larva has made its first moult, when it begins to eat diatoms. Over the course of the next five instars, food is stored in the body as fat droplets that accumulate to become clearly visible by the final nauplius stage, when it moults into a cyprid larva. This is a completely different-looking organism, more like a bivalve, with two transparent valves, six legs and a pair of antennules that have cement glands inside. As the cyprid is non-feeding, the accumulated fat provides the fuel to enable the larva to search for a suitable resting place. At that point the cement is released to glue it permanently in place as it moults into an adult barnacle. Copepods also have a nauplius larva for six instars, but as the adult remains in the plankton the final nauplius moult produces a juvenile adult that undergoes further moulting to reach full size.

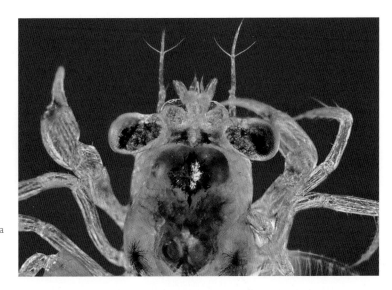

ABOVE: The megalopa larva of a crab, which will soon settle out of the plankton and develop into an adult.

Having a mobile larva in the plankton is also a useful way of meeting the opposite sex, and some quite exceptional and complex methods for doing so have evolved. Epitoky in the syllid worms was described in Chapter 6; in essence, this is where the polychaete worm has no gonads and produces a chain of up to five stolons (specialised worms for sexual reproduction) at the end of the body. They can be male or female, and by what appears to be a rather explosive technique they are released from the adult to drift up into the plankton. They do not feed, but after mating the male dies and the female stolon cares for the eggs until she perishes. Another polychaete worm, the Sand

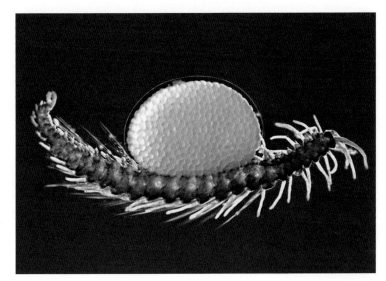

Epitoky: a female stolon of a syllid worm protects the developing eggs while in the plankton.

Mason, goes through two phases in the plankton during its life cycle. It has the normal trochophore larva, which settles to produce a worm on the shore but then returns to the plankton in a phase called the aulophora, a tentacled worm living in a tube. After several stages it drops to the shore to become a sedentary adult. The variety of shape and form in zooplankton is quite extraordinary.

Supply-side ecology

On the night of 15 February 1996 the 150,000-tonne tanker *Sea Empress* struck the rocks in the entrance to Milford Haven, and we thought the end of the world as we knew it had occurred. Over the ensuing days, 72,000 tonnes of Forties Blend crude oil leaked from the vessel and covered local beaches. The seconds and minutes were marked by the sound of shells falling off the backs of dying limpets onto the rocks below. The shell-less animals then slowly expired from dehydration, and some local shores had more than 95 per cent mortality of limpets. Fortunately, as this occurred so early in the year, plankton blooms were unaffected and occurred as normal in the Atlantic and Irish Sea. Adult molluscs elsewhere along the Welsh and English coast produced their normal autumnal larval peak, and the sea currents brought veliger larvae to settle on the now clean and empty rocky shore. The balance of molluscs may have been slightly different, but the sea supplied sufficient larvae to allow the process of recolonisation and recovery. The oil had wiped out many generations of limpets, but they were replaced by juveniles all of a similar age.

In this way, the ocean currents and tidal action mix and disperse the plankton. Larval stages are brought to the shore and their survival depends on the coincidence of favourable conditions. So if a cyprid of the common acorn barnacle *Semibalanus balanoides* arrives on the upper shore in early summer it is likely to die of desiccation as the tide drops. If the cyprid had found the middle shore it would have had a higher chance of surviving the abiotic conditions there. But all larvae have a limited life span, and so if the sea cannot deliver cyprids to the shore in time then the barnacle population will not be replenished. This is the basis of supply-side ecology. Life on any rocky shore is dependent on the larval stages that arrive from the plankton and whether they are able to find the appropriate range of conditions within which they can survive.

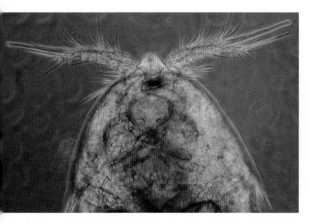

Close-up of the head of a copepod, showing the single median eye retained from the nauplius larval stage.

Permanency, or holoplankton

The most abundant multicellular animals swimming in the sea are copepods. This group forms a major component of the holoplankton, with probably more species than any other crustacean group. They are a common sight in a plankton sample, using their muscles to vibrate and push their way through a packed cluster of diatoms and detritus, feeding on them as they go. Copepod mouthparts create a continuous flow of water to collect food, which is diverted into the mouth. Estimates of diatom consumption by a single large *Calanus finmarchicus* can exceed a quarter of a million individuals a day. However, these copepods are not obligate vegetarians; they are happy to eat animals if they will fit in their mouths. In fact, limited mouth size may be the reason that over half of the known copepods are parasitic forms, found on a very wide range of animals including fish.

Another crustacean group found in the holoplankton is the Cladocera or water fleas. Elsewhere they include species such as *Daphnia*, common in ponds and lakes, but marine varieties are not so diverse. *Evadne nordmanni* is the species most likely to be seen around the British Isles, where it is an important predator in the spring plankton.

Rotifers are a well-described group of animals found in freshwater plankton, but less well known are the marine rotifers. They are common members of the holoplankton along the coast. Bdelloid rotifers have been extensively studied across the world, but not a single male has ever been found, just sexual females. Outside of the rotifer group occasional examples of this are found – such as in aphids – but nowhere else in an entire class of organisms.

While on the subject of uniqueness, we should mention the arrow worms. Just visible to the naked eye, individuals of the genus *Sagitta* occasionally appear in the permanent plankton along our Atlantic coasts, although they are more typically found as offshore, oceanic animals. Little understood, they are in a phylum by themselves called the Chaetognatha, a name that reflects their bristly mouthparts. These help them detect vibrations in the water caused by copepods, which they grab with hooks after immobilisation with a neurotoxin.

Plankton: drifters of the shore

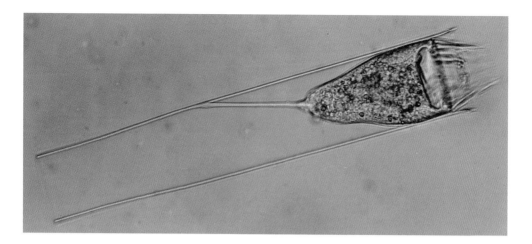

Firm favourites of Ernst Haeckel were the beautiful radiolarians, which he drew in exquisite detail. Books of his work are still available, with the most recent publication compiled by Olaf Breidbach in 2005. Radiolarians represent species from the protists. These single-celled organisms are widespread but not always abundant in plankton samples found along the rocky shore. Tintinnids, on the other hand, regularly turn up in in-shore samples; they form the biggest group of ciliate protists worldwide and are almost entirely marine. The characteristic feature is a wineglass- or bell-shaped covering, or test, made of chitin and protein called a lorica, which is normally crystal-clear so that the cell contents are visible inside. The anterior end of the lorica is open, often with a collar, and has a row of cilia. Tintinnids can be quite mesmerising to watch but do not survive long in collected samples, and it is more usual to find the empty lorica appearing under the microscope.

Tintinnids form the largest group of marine ciliates. This is a *Eutintinnus* species.

* * *

Plankton varies both seasonally and spatially. Our understanding of this ever-changing environment is so important that it is monitored, globally, by a number of organisations. Phytoplankton, in particular, plays a critical role in the carbon cycle and is responsible for producing over half of the earth's oxygen. This chapter has considered just the narrow strip of coastal sea around the British Isles, but it demonstrates the huge complexity of what drifts onto and past our rocky shores. The relevance of healthy plankton communities to the wellbeing of rocky-shore organisms – and indeed human welfare – cannot be overstated. The biodiversity of our shores is completely dependent on the twice-daily delivery of plankton brought in by the tides.

Attack from air and sea

chapter eleven

Mammals, birds and fish are important top predators of the shore. In most cases, they are day-trippers. Despite some major adaptations, they are unable to remain permanently within the environment supplying their food – and so, with the rise and fall of the tide, they must periodically move out, to continue feeding elsewhere, to breed, or simply to rest. These animals have an ecological impact on the shore, but they also transfer a high level of nutrients into and away from the habitat. In this respect, as in so many others, the seashore functions differently from other ecosystems.

Spring is a time of great anticipation on the rocky shore – a season of coastal change. Plankton blooms, the spawning of many shore organisms and fish arrivals herald a shift in the marine ecosystem. Shrill calls announce the arrival of the 'sea swallows' or Arctic Terns *Sterna paradisaea* that have returned from Antarctica – and are just passing through on the longest migration known of any animal. Flitting and hovering around the water's edge at low tide, they search for small fish that have moved down with the water after feeding on the shore. Working in the offices at Dale Fort, as Easter approached our glance would frequently be drawn to the window, looking out for the terns or the first Gannet of the season.

Busy cliffs

Summer survey work beneath the cliffs of the Skomer Marine Conservation Zone would lack a little something without the company of Kittiwake cries and Guillemot growls. Seabirds appear as if from nowhere in the early part of the year, establish (or re-establish) territories, acquaint (or re-acquaint) themselves with mates, copulate, lay an egg

An Oystercatcher feeding on shellfish in the middle shore.

(or eggs) and ultimately make the best of bringing up fledglings. In the process they create an incredible mess with their guano. Non-resident birds – including the auks (Puffin, Guillemot and Razorbill *Alca torda*), Gannet, shearwaters and petrels, terns and Kittiwakes – then depart, some as early as July, some as late as October, and the cliffs and stacks fall eerily silent until the following spring. For much of the year, these birds live far out to sea, some even as far away as the coasts of Argentina and Brazil. However, to breed they need solid ground on which to lay eggs. Adaptations that allow the successful conquest of the sea, such as supreme diving and swimming ability, mean that life on land can be a struggle. Auks have small, narrow wings, superb for swimming but not for rapid take-off, and landing is barely more than a controlled crash. Shearwaters have legs at the back of their bodies; when they land they flop forward onto their chests and are extremely vulnerable to aerial attack by Great Black-backed Gulls *Larus marinus*.

Ground predators such as Red Foxes *Vulpes vulpes*, Brown Rats *Rattus norvegicus* and feral cats pose the biggest problem for nesting seabirds. Cats, in particular, often introduced by lighthouse keepers, have been known to devastate entire ground-nesting populations. To avoid these predators, birds such as the Herring Gull *Larus argentatus* and Lesser Black-backed Gull *L. fuscus* find a rock stack, islet or island with flat, grassy areas above the splash zone to nest on. Steep cliffs with inaccessible rocky ledges are the answer for species such as Guillemot and Razorbill, but these high tenements quickly fill up as the birds compete for space. Kittiwakes also favour steep cliffs, but unlike the Guillemot and Razorbill they construct a neat nest to hold their precious clutch of eggs. Puffins opt for burrows near the top of the cliff, while Cormorants *Phalacrocorax carbo* and Shags *P. aristotelis*, unlike many other seabirds, produce large nests of sticks and seaweed, which require a lot of space. An ill-advised choice of nest site, on the rocky base of a cliff dangerously close to the splash zone of high-energy seas, can cause their occasional demise.

The gulls are shore predators but will also cannibalise one another's chicks if they are left unattended. The Great Black-backed Gull is the top coastal predator, with Rabbit *Oryctolagus cuniculus*, Puffin and Manx Shearwater *Puffinus puffinus* on the menu if available. This rather selective diet is supplemented by general scavenging for any suitably sized prey. We recently observed an individual steal an octopus from a rather bewildered Cormorant and proceed to attempt to down it in one. Tentacles, suckers and writhing arms made this rather challenging,

A group of Shags on the upper shore.

and we were entertained for nearly half an hour before the octopus succumbed. Puffins are significantly quicker to eat and can be swallowed in one go. Adult Rabbits are a bit more feisty; we have seen Great Black-backs catch and air-lift them to a suitable height before dropping them on the rocks below to kill them for easier consumption.

Great Black-backs, along with the other gulls, regularly feed and scavenge on the shore at low tide for starfish, crabs and any other organic miscellany that takes their fancy. During non-breeding periods, however, they may move around the coast and disappear completely from shore areas, while some such as the Herring Gull move inland to scavenge and roost at all times of the year.

The Herring Gull is an important scavenger on the shore.

Cormorants are another common sight, fishing along the edge of the rocky shore. Their feathers lack waterproofing, an interesting adaptation to allow rapid diving for food: as the plumage takes on water this drives out air that would otherwise make them buoyant. They even lack internal air bags, and their bones are heavier than those of most other birds – the extra weight helps with diving. Such efficient fishing means they spend only a short time in the water and the rest roosting above the waterline to dry out. Wings are spread wide open to encourage rapid drying; this exposes a large surface area of black plumage to absorb heat, enabling faster digestion of large fish.

Although an integral component of the coastal ecosystem, not all these seabirds feed on or even near the shore. Cormorants and shags may catch fish close inshore, but the auks are deep divers, feeding on sand eels, Sprats *Sprattus sprattus* and other small fishes in the water column; Guillemots dive as deep as 200m for their food. Kittiwakes, terns, shearwaters and Fulmars, on the other hand, are essentially surface feeders but will dive a metre or so when necessary (even Gannets, with their famed plunge-diving, do not venture more than a handful of metres below the surface). These birds' contribution to the shore ecosystem is through nutrient enrichment, which stimulates plankton productivity, and thus the shore can act as a nursery and feeding site for seabird prey.

The guano effect

On a clear day, the island of Grassholm can be spied out on the horizon from mainland Pembrokeshire. Even in winter, when Gannets have deserted the rock, the upper portion stands out white against the blue skyline. Over many years, guano deposited by the birds has accumulated around their nesting sites, painting the rocks in a pale excrement that is rich in nitrogen, phosphorus and potassium. Once, in the days before the Dale Fort boat had radar, in thick fog that had descended without warning, we found our way to Grassholm by smell alone.

This excess of nutrients has an impact on surrounding ecosystems. For one thing, the plants around the nesting site become dominated by a few halophytes (species adapted to high levels of salt in the soil). Rain washes nutrients into the sea from the cliffs above, where they act as a fertiliser on the diatoms in the coastal plankton, increasing the chance of localised phytoplankton blooms. Steep seabird cliffs with deep seawater below may have an altered community composition in the sublittoral region owing to the nutrients descending from above and attracting a varied range of animals that exploit them. A good example is on Skomer, below high cliffs at the Wick, where the rocks are covered in the rare Scarlet-and-Gold Cup-coral *Balanophyllia regia*.

Rocky shores directly below seabird cliffs show a reduction in clear biological zoning, with a loss of some *Fucus* species. Perhaps most interesting is that the orange splash-zone lichens are able to

Part of the Gannet colony on Little Skellig Rock, County Kerry. The green colour on the rock is the alga *Prasiola*, which is tolerant of very high nitrogen levels. Note the splash-zone lichens high up the cliff.

grow significantly higher up the cliffs as a result, although this seems to be limited to where seabird colonies are in exposed places. The presence of guano may also stimulate increased abundance of green algae – *Prasiola stipitata* for example. Although small and flat, this seaweed can cover large areas, producing a green slime when wet, and is tolerant of high levels of salt, other minerals and freshwater runoff. When *Prasiola* is abundant in the splash zone and upper shore, other species rarely become established. In the middle shore, with a reduction in the wracks, red seaweeds including Grape Pip Weed and encrusting species such as Hildenbrand's red weed *Hildenbrandia* spp. may increase. In general, enrichment of the rocky shore by guano-related nutrients reduces biodiversity to the benefit of a few species, whose biomass increases.

Guano is an ideal fertiliser for use in agriculture. Fortunately the British Isles has escaped intensive commercial exploitation of seabird colonies for this product (although colonies have been spectacularly overexploited for their flesh, feathers and eggs). Elsewhere in the world, guano trade was big business. In the early 19th century the UK imported over two million tonnes of guano from Peruvian seabird stacks, and in 1867 US Secretary of State William H. Seward agreed to purchase Alaska from Russia for a little over seven million dollars. The reason? – to tap into its rich guano resources. The process of collection was incredibly destructive: it could involve scraping a 10-metre depth off the top of a seabird island colony, totally destroying the birds' habitat in the process. Mercifully, as elsewhere around the world, there is little evidence that guano was collected by local communities on remote Alaskan islands, since seaweed was the preferred fertiliser.

An excellent account of the lives, loves and behaviours of seabird populations and the issues they face today can be found in *The Seabird's Cry* (Nicolson 2017).

Birds feeding on rocky shores

So far, many of the birds we have discussed use the coast for breeding and then return to the open ocean, and thus might be described as true seabirds – though Shags, Cormorants and some of the gulls are more difficult to pigeonhole in this way. However, a number of other bird species stay as permanent residents on the rocky shore, including the strikingly coloured Oystercatcher, with its characteristic black and white plumage contrasting beautifully with the deep orange-red of the legs and bill. Found singularly or in small groups, it feeds with the descending tide on a wide invertebrate diet that ranges from mussels and limpets to small crustaceans and shrimps. The shape of the bill can give some indication of an individual's preference, as wear and tear to the structure is repaired by regrowth: regular consumption of soft food encourages a long, slim bill to form, while a diet involving breaking into hard shells produces a tough, thicker 'hammer' bill. When disturbed, Oystercatchers emit a distinct, piercing call and have a tendency to take flight, moving parallel to the shore and very close to the water surface to shift away from the perceived threat before quickly settling to continue feeding. We once inadvertently disturbed a bird in the Outer Hebrides that had been lying still on some beached kelp, but rather than flying away it slipped into the water and dived,

reappearing some distance away and swimming back to some rocks before taking flight. This amazingly adaptable species is found on most rocky shores: at high tide they roost in the splash zone or find a stretch of grass, where they typically search for worms or sandhoppers while waiting for the tide to drop again. If the shore is relatively free of people, Oystercatchers will nest at the top of the splash zone, making a scrape among the lichen and Thrift, adding dry seaweed and small sticks.

Other waders frequent the rocky shore but are not necessarily permanent residents; they may breed elsewhere (often on moorland) and retreat to the coast for the winter. Turnstones can be found on almost any type of shore, flipping over debris and seaweed looking for sandhoppers. They breed in the high Arctic, but small groups can be seen around our shores in any month of the year. Their plumage is an irregular pattern of black, brown, yellow and green over a body of white, giving them the appearance of animated pieces of seaweed or lichen. Two other small, plump waders, Purple Sandpiper *Calidris maritima* and Dunlin *C. alpina*, appear in small numbers to feed on rocky shores, especially during winter, although they are usually only conspicuous when they take flight. Amongst the non-wader species, Rock Pipits are significant predators on rocky shores (see p. 322), and Grey Herons *Ardea cinerea* are often seen around pools, collecting crabs and fish.

ABOVE: Grey Herons are very territorial, and are commonly seen feeding on sheltered shores.

Rock Pipits (inset) are often found in the splash zone feeding on seaweed flies, fly larvae, isopods and sandhoppers.

Fish

Fish are an important part of the diet of birds and mammals, but compared with the open sea there is a real paucity of species associated with the rocky shore. This is not surprising, considering that fish have mostly evolved to be truly aquatic swimmers with fins and a tail for locomotion, and gills to obtain dissolved oxygen from the water. Bony fishes (as opposed to the cartilaginous ones such as shark and skate) have a swim bladder above their gut, running the length of the body. Acting as a buoyancy aid, this helps to maintain the position of the fish in the water column without the constant need for swimming – not especially helpful for life on the shore with an ever-changing state of the tide. Ultimately, this has led to fish becoming adapted either to moving in and out with the seawater, or to surviving emersion.

Except for some very young stages in their life cycle, the wrasse family are members of the 'move in and out' group, and they have a typical fish-like appearance. Wrasse develop an amazing show of colour, especially during the breeding season, but perhaps their best party piece is the ability to change sex, usually from female to male. The Ballan Wrasse *Labrus bergylta* is a carnivore living in the sublittoral zone that moves with the tide and feeds on a wide range of crustaceans including most species of crab, spider crab and prawns.

Corkwing Wrasse move in to the shore during high tide.

Cormorants are active hunters of rocky-shore fish. This individual has caught a Fifteen-spined Stickleback, usually found in deep rock pools.

Its teeth enable it to crunch mussels and scrape barnacles from the rocks. Ballan Wrasse tend to be inconspicuous, but you may be able to spot a nest consisting of a crevice lined with coral weed and other softer seaweeds in deep pools at the bottom of the shore. Eggs are laid in the nest and fertilised by the male, and then deserted. On hatching, the youngsters remain in the pools, feeding on small crustaceans, before eventually leaving for deeper water. While Ballan Wrasse are the commonest species of wrasse found along the rocky coast, relatives such as Corkwing Wrasse *Symphodus melops* and Cuckoo Wrasse *Labrus mixtus* also appear now and again.

While the wrasse are the most distinctive of visiting fishes living close to the shore, there are plenty of other day-trippers that move in with the tide to find food. One worthy of special mention, as there can be odd years when a mass arrival occurs, is the Lumpsucker *Cyclopterus lumpus*. This species lives at the extreme edge of the rocky shore, feeding on crustaceans and other fish during high tide and disappearing as the water recedes. It has a small mouth and eyes, and a chunky body that appears to have rows of rivets holding it together; these are actually bony plates and tubercles. Its shape helps it to cope with a remarkable range of wave action, and its pelvic fins are modified to act as suckers and hold the fish tightly to the rocks. Clearly these adaptations aren't always successful, however, since a late spring gale at Kimmeridge Bay in 1986 resulted in many hundreds of bright male Lumpsuckers being killed and stranded at the top of the shore.

Lumpsuckers are unmistakable in breeding fettle, when the normal greyish colour of the males turns a remarkably bright red. Both sexes can grow to almost half a metre in length. They spawn and lay colourful blue-red strings of eggs in the sublittoral and lower shore, where the male provides parental care for a few weeks, fanning the eggs with fresh seawater. The resulting small juveniles are green, and can survive under large stones in the lower shore.

For a fish to remain on the shore throughout the tidal cycle, major adaptations are required for survival, and there are just a handful of species in the British Isles that are capable of this. The Shanny or Blenny *Lipophrys pholis* – first introduced in Chapter 3 – is a fish that has truly made the rocky shore its home. For one thing it has lost the swim bladder, which would otherwise keep it buoyant. Even in a rock pool this fish keeps low in the water and drops instantly to the bottom the moment it stops swimming. Juveniles live in shallow rock pools almost anywhere on the shore. Walk slowly across any relatively exposed beach and you will notice the young fish darting around in the pools as they become aware of potential danger. A pool half a metre square is likely to have one or two youngsters, each with a distinct territory that will be assiduously fought over. At this stage, the young eat the cirri of barnacles. As they grow, their former territories are relinquished and they move into larger pools.

Fully grown Shannies can reach 16cm in length and often feed on crabs and other crustaceans under rocks away from their territorial pool. It is here that they rest when the tide goes out, and consequently they rely on adaptations to help them breathe in open air. Their gills have strengthened arches to prevent them collapsing out of water; if this were to happen they would lose all their surface area for gas absorption, which would cause asphyxiation. As long as Shannies limit their movement, their oxygen demand will not be high. Turning over large stones in the middle shore will reveal plenty of variation in colour of Shannies, owing to the need for camouflage among different

An adult Shanny will eat a wide variety of food, from seaweed to barnacles and crabs.

seaweeds and pools, and placing a Shanny in differently coloured buckets can actually induce some changes in its appearance: most obviously, it may become darker or paler in hue.

Close relatives of the Shanny that can also turn up on the shore include the Tompot Blenny *Parablennius gattorugine*, found in the very lowest part of the shore, and Montagu's Blenny *Coryphoblennius galerita*, found in the middle and lower shore but only in southwestern Britain or on the Atlantic coast of Ireland. Both of these species have a pair of branched tentacles above the eye. You may like to follow the blog of a Tompot Blenny, www.bennytheblenny.com, the 'fish with antlers'.

While the Shanny is widespread and common on any rocky shore, other species are less prolific. The nearest in abundance might be one of the rockling species, especially the Five-bearded Rockling *Ciliata mustela*. These are quite different in appearance, almost like small eels. Their eyes are much smaller than the Shanny's, and it is thought the 'beard' of barbels on the front of the head is used in dim light for locating prey, mainly crustaceans. Rocklings occur in pools and under stones, and although the Five-bearded Rockling can be found all around the British Isles, it is not always common on the shore.

Similar in habit and distribution to the Five-bearded Rockling is another interesting species, the Butterfish. On some shores they can be a common occurrence, while on others they are scarce or non-existent. Their distinctive bodies are long, slightly flattened

The small Shore Clingfish has a flat body and lives under rocks where it can firmly attach with a sucker adapted from a pelvic fin.

and slender with a long thin dorsal fin, which has large black spots. Butterfish are distinctly slippery when you try to pick them up, and indeed this is how they get their common name. They are unusual fish in that both sexes take turns to guard the eggs by curling their bodies protectively around them to prevent accidental dispersal.

You are unlikely to find seahorses on most British or Irish rocky shores, but you may come across some of their relatives, the pipefishes. Like seahorses, pipefishes lack scales, and instead have a series of jointed bone-like rings encircling their bodies, giving them a rather rigid appearance. Worm Pipefish *Nerophis lumbriciformis* and Greater Pipefish are common in the south and west on sheltered shores amongst seaweed, where they feed on small crustaceans. They do not make rapid movements, just a slow uncoiling when disturbed in the seaweed, so can easily be overlooked. After a female has laid her eggs the male carries them within a groove under his body, all neatly attached in rows. Once they hatch, like most male fish on the shore, he will look after them.

In the increasingly stressed seabird colonies to the northeast of Britain, Puffins are bringing back pipefishes to feed their young instead of the preferred fare of sand eels. This is disastrous, as pipefishes have a far lower fat/oil content, essential for successful puffling growth, and they are bony; sometimes the young bird chokes on the parental offering.

A young Topknot: the only flatfish to be found on the rocky shore, in lower-shore pools. It lies on its right side and has dark blotches across the eyes.

Clingfishes also occur on shores in the south and west, although they are more difficult to find. A sucker under the body, which is a modified pelvic fin, gives them a strong grip on the rocks – from which they can be difficult to remove. The Shore Clingfish *Lepadogaster purpurea* is red-brown with a blue spot behind each eye. Like many of the fish that can be found here, it is carnivorous, feeding on invertebrates. It favours sheltered shores and has a fairly flat body so that it can live under rocks. The only true flatfishes that occur on rocky shores are juvenile Topknot *Zeugopterus punctatus*.

Searching for fish on the rocky shore can certainly result in surprising discoveries – finding a beautiful Lesser or Curled Octopus *Eledone cirrhosa* in a lower-shore pool on Baleshare Island in the Outer Hebrides is a particularly memorable one for us. Associated more with the sublittoral, this species will come on to the shore to feed. It is around 30cm in length, so quite small for an octopus, and the tentacles, with a single row of suckers, curl at the end when at rest: hence the common name. Found in the English Channel, Atlantic and North Sea, it is becoming a widespread species, possibly owing to a decline in predators such as Atlantic Cod. In turn, the octopus could be considered a pest by the fishing industry, given its fondness for eating crabs and entering lobster pots to feed.

Blennies and the prawn *Hippolyte varians* are good at changing colour, but the real chameleons of the shore are the cuttlefishes. They can quickly change both skin colour and texture, and this is used for

Common Cuttlefish feed on small crabs and shrimps, manoeuvring slowly with a rippling action of their side fins.

communication between individuals as well as for camouflage, and for deterring predators such as seals and seabirds. Common Cuttlefish *Sepia officinalis* normally appear as washed-up white cuttlebones in the strandline, but live animals are occasional visitors to the shore, and there are instances when individuals become trapped in deep pools at low tide. Cuttlefishes can use their tentacles to catch fast-moving prey such as prawns and fish, but they can also make short work of aggressive crabs by using their powerful arms and razor-sharp beak. If confronted by a crab impressively waving its claws in defensive posture, a cuttlefish will simply avoid the problem by attacking from the rear. As well as feeding on the shore, Common Cuttlefish lay their black grape-like eggs onto seaweed in the lower shore.

The Little Cuttlefish *Sepiola atlantica* is a delight to find on the shore but is associated more with sediment, and is found in the Channel, up the east coast to Scotland and very occasionally in Wales – but unfortunately sightings are quite rare. It is around 50mm long with rounded fins on the side of the body and can be transparent, although its chromatophores also allow for rapid changes in colour.

Octopus and cuttlefish are complex molluscs, belonging to the class Cephalopoda. They have well-developed nervous systems and good eyesight, and their blue blood contains haemocyanin, a pigment based on copper. Of all the animals that might be encountered on the shore, these are some of the most intelligent, and tend to give the impression that they are studying you, rather than the other way round.

Silkies, cetaceans and other mammals

In 2001, the foot and mouth crisis in the British countryside prevented access to all but a few coastal locations, which had the benefit of encouraging us to visit sites that we did not often frequent, such as Stackpole Head, Pembrokeshire. On one occasion as we sat down with a group of students to eat lunch, three huge dorsal fins suddenly appeared out of the water in front of us, barely 50m from the low tide mark. The distinct dorsal fins were followed by the brief appearance of black and white bodies, and then the Killer Whales *Orcinus orca* were gone.

Further north, Cardigan Bay is a designated Special Area of Conservation (SAC) because of the cetaceans that occur around the west coast of Wales. Porpoises and dolphins can be seen coming in close to feed at points where upwelling nutrients feed their fish prey off headlands such as Wooltack Point and Strumble Head. The Moray Firth in Scotland and the west coast of Ireland around Galway Bay and Kerry are equally good regions to view these iconic mammals. Killer Whales tend to be seen hunting offshore near islands where seabirds breed. Our sighting was at Easter, a time when seabirds are returning to their nests, and this may be why they were so close to land, but they also like seal meat, which could have been the attraction on that particular day.

Two species of seal occur in British waters: the Common Seal *Phoca vitulina* and the Grey or Atlantic Seal *Halichoerus grypus*. The former

A Grey Seal swimming amongst the kelp.

is the smaller of the two and is associated with sandy shores and the eastern side of Britain. The second species grows to over 2m in the case of the bulls, and they can often be seen hauling themselves up onto the rocks at low tide, bellies full of fish and cuttlefish, to bask in the sun. If there is any disturbance they will quickly flop back into the sea. The Grey Seal can appear anywhere around our rocky coast, and individuals travel far and wide; Skomer seals form part of a population that encompasses the whole area from the Irish Sea to southwest Britain's southern approaches.

Female Grey Seals (cows) tend to return to their birth area, and will come to rocky and shingle beaches to have their young. In Cornwall and north to Pembrokeshire, the first pups appear on the shore around late August, rising to a peak in October, though in Scotland pups may not be born until November or December. After giving birth, the cows return to the sea but regularly come back to the shore to suckle their pups with a very rich milk that helps them bulk up as fat accumulates. At three weeks the pups are abandoned by their mothers, who around this time will have mated with a bull and are moving away from the birthing shore. The gestation period appears

This Grey Seal has just given birth on the shore (the afterbirth is still visible). The newborn pup is quite slim and covered in long, pale yellow fur.

to be a year but is actually only nine months; the embryo stops developing for three months over the winter. This delayed implantation helps to ensure that seal pups are produced at roughly the same time each year. Meanwhile the pups, having been left without food, are forced to fend for themselves, hunting along the shore for crabs and other easily captured invertebrates. When they are strong enough, the young seals from the Welsh and southwest populations head towards France, while the Scottish youngsters are likely to move north to Norway.

A two- or three-week-old Grey Seal pup.

Traditional hunting of seals and Gannets in Scotland

I am a man upon the land
I am a silkie in the sea
And when I'm far from every strand
My home it is in Sule Skerry

These lines are taken from *The Great Silkie of Sule Skerry*, a traditional folk song from Orkney that was not known outside of the isles until the mid-19th century. In the song, a woman is explaining that she does not know the father of her son, when a man appears, claiming to be a silkie (a man on land and a seal in the sea). He takes the son and gives the mother a purse of gold, saying that he is likely to be shot on Sule Skerry (a remote island 60km west of Orkney), a reference to the men that annually went to the rocky islets to shoot and return with several hundred seals. Grey Seals were hunted almost to extinction in the UK and became the first mammal to be protected by law: the Grey Seals Protection Act of 1914.

The seal shoot was not unlike the tradition of catching Gannets on Sula Sgeir. This small island of hard gneiss rock is even more remote, lying 80km further west than Sule Skerry, 60km from the Butt of Lewis. Each year in autumn a group of up to a dozen men would set out for the annual collection of up to 2,000 young Gannets, known locally as *gugas*. Today they need a special licence from Scottish Natural Heritage to make the collection, which is controversial not least because of the method of using a pole and rope noose to grasp the birds. Guga meat is highly prized and sought after, although it is very much an acquired taste. The original scientific name of the gannet was *Sula*, possibly from Sula Sgeir (or it may be that the island is named after the bird).

An Otter hiding among middle-shore Egg Wrack.

Otters are a conservation success story, with a steady increase in numbers over the last few decades both in rivers and estuaries and along the coast. More than half of the Scottish Otter population live exclusively on rocky shores, which are magical places to catch a glimpse of these elusive creatures, as here they are active by day. Feeding on crabs, urchins, shellfish and fish, they often leave skeletal remains such as legs, bones and urchin tests on top of the wrack. Thankfully this is now an increasingly frequent sight on the rocky shores in Wales and parts of the West Country too.

The abundance of shore food means that the territories of coastal Otters are smaller than those living in freshwater. The only downside to their seafaring existence is that the salt compromises the insulation power of their fur.

* * *

Attack from air and sea

More than any other habitat, perhaps, the rocky shore is inextricably linked to the environments that surround it. It offers a place of safety for birds that may find their food inland or far offshore, while for many other organisms the shore itself functions as a food store, open every day of the year to whatever animal wishes to visit and feed (humans included). In winter it may be the only place available for a deer to find vegetable matter, or for a fox to crunch crabs. The deer and fox, in this way, will be moving organic matter to other environments. Fishes, visiting the shore to feed, likewise move nutrients away into deeper water, while the ocean-feeding seabirds shift them in the opposite direction. For the rocky-shore ecosystem to succeed, there is a need for an input and recycling of nutrients. To understand the story of nutrient recycling fully, however, we need to visit the top of the shore and examine the strandline in some detail.

Over two million people are believed to be sea anglers in the British Isles, with over three million sea-angling days a year in England alone (DEFRA survey report 2013). This makes humans a significant top predator of the shore.

Nature's giant compost heap

chapter twelve

Beachcombing along the strandline can be a rewarding pursuit, particularly after a gale has washed in a lot of debris. Rummaging through the material deposited at the top of the shore is easiest on sandy or shingle beaches, and it is perhaps tempting to think that these are the only places where strandlines occur. But all shores, including rocky ones, have these transient lines of material, both natural and of human origin, that the sea leaves behind as it recedes. These are compost heaps of mammoth proportions where nature recycles essential nutrients, not only back into the marine ecosystem but also into the terrestrial. At this junction of two environments, many land-based animals come to reap the rewards of the marine bonanza.

Strandlines can be full of surprises. We had been working in the Outer Hebrides in the summer of 1977 when reports started coming in of a large mass floating down the coast of South Uist. The bulge in the shore's strandline was clear from more than a mile away and, as we approached, the smell was quite distinct as well; it turned out to be a Sperm Whale *Physeter macrocephalus* that had been dead for many weeks, brought ashore in a gale along with a considerable amount of seaweed.

A strandline cannot be deemed a 'zone' as such, since it changes so much with the tide and season. On a rocky shore, furthermore, the debris may be scattered as it gets caught in upper-shore pools or attached to seaweed. But the strandline, however ill-defined, is important. What makes the composting that takes place here different to any other is the immense biomass involved and, more importantly, the speed of the recycling. Decay and odour pervade the scene – and could even perhaps explain the general lack of research on strandline ecology.

OPPOSITE PAGE:
Stranded seaweed, mainly kelp, after a storm in the North Sea.

A dead Sperm Whale on a South Uist beach among the stranded seaweed, 1977.

As the Sperm Whale demonstrates, strandlines can give us important clues to what lives offshore. Other more commonly encountered items might include 'mermaid's purses', which are the egg cases of dogfish and skates, and cuttle bones of cuttlefish. Over the centuries, human activity has changed the composition of the strandline, and in the last 50 years we have seen an exponential growth in plastic, polystyrene and other waste.

Strandlines are dynamic zones on the shore, changing with every tide, and a shore may even have several strandlines (see image opposite). Spring tides create a band high on the shore, and successive tides through to neaps may form smaller lines of debris culminating in a larger neap strandline. The composition of each can vary along with the animals that quickly colonise them.

'Mermaid's purses' are the empty egg cases of cartilaginous fish such as dogfish and other sharks – a common find in the strandline.

Compost composition – dead organic matter

In Chapter 5 we discussed the beauty and rapid growth of seaweeds, the 'plants' of the shore. The irony is that, despite the enormous productivity, very little of the photosynthesised material is actually eaten by the shoreline animals. When the seaweeds die and decay, however, their nutritional content becomes more accessible. Large, complex molecules break down and become smaller and simpler for organisms to ingest and utilise. Bulky organic matter such as kelp decays into small or microscopic particles that drift easily in the water, along with any accompanying dead animals or exoskeletons that have been shed by their previous owners. This material is the important particulate detritus that dominates the seawater around our coasts, and it is essential food for filter and suspension feeders such as barnacles, sponges, sea-mats, and sedentary worms.

The way in which this material is released into the environment varies. Seaweeds may be ripped whole from their anchorage by storms at any time of the year, and wear and tear to the fronds occurs constantly. In addition, kelp blades may be shed in their entirety in the autumn or spring as part of the natural growth cycle (see Chapter 5). Throughout the year, therefore, nutrients will diffuse

Different tidal strandlines: the upper, spring-tide line consists of what is left of decayed and now dry kelp, while fresh material has been deposited by a more recent neap tide.

from seaweeds into the surrounding water, but the most notable times for strandline build-up are after storms, when the kelp and wracks may form dense heaps up to several metres high. Spring and autumn are the peak periods.

When wracks – especially Egg and Bladder Wrack – become detached in summer, they can form living floating rafts before they become stranded. Additional seaweeds such as the rapidly growing *Ulva* species can then become entangled in these masses, which fish and crustaceans temporarily colonise, taking shelter and feeding in the open water. While the material is floating, decay is slow to begin; seaweeds have no roots, obtaining all the nutrients they need from seawater through the frond surface, and so can remain viable while drifting. Once stranded, however, decay begins in earnest – and it will include the unfortunate animals still trapped within the raft.

Seaweed rafts can travel great distances. The Gulf Stream brings debris from the Caribbean, Central and South America, and Coconuts *Cocos nucifera* and large beans can appear on the exposed west coast of the British Isles after crossing the Atlantic. Sea beans come from the tropical *Mucuna* lianas or vines; their large seed pods may develop on plants that overhang cliffs, with the result that the beans drop into the sea and may be carried in the currents. Initially they can get caught up in vegetation rafts, but gases inside make them buoyant, enabling them to float unaided for thousands of kilometres.

This Coconut has been carried in the Gulf Stream and deposited on a Dorset beach.

Rocky-shore microorganisms – bacteria and fungi

Bacteria are abundant in our coastal seawater and across the rocky shore. They are still present further out into the ocean but less dense in number and variety. While a few species are dangerous parasitic forms (for example on fish stocks), the vast majority are saprophytic (feeding on dead plant matter) and collect around the particulate detritus suspended in the sea.

Fungi are also important saprophytes of the coast and sea but are not well documented, particularly when it comes to understanding their role in the marine environment, and they receive scant comment in marine books. Currently, there are fewer than 500 marine species known worldwide, which is barely 1 per cent of the total number of fungi. The other 99 per cent are terrestrial, and this number is growing regularly as new species are described each year. By contrast, very few new species of marine fungus have been added to the list since 1980.

Much of the open ocean is akin to a desert, with low biodiversity leading to poor productivity and recycling of nutrients. It is only coastal areas and upwelling zones that buck this trend, courtesy of the nutrient input into these areas. Similarly, offshore fungal species are limited to a few yeast-like examples. Most marine fungi are filamentous forms associated with the rocky shore, and more than a third live on the surface of seaweeds. The majority of the remaining fungi can be

A marine fungus growing on the red alga Dulse. Notoriously difficult to identify, it is probably a type of *Saprolegnia*.

found in the strandline. The paucity of marine species of fungi does not seem to be related to salinity, but rather, perhaps, to the availability of oxygen. The whole subject begs for more research to be done. Some of the genera most associated with seaweeds include *Penicillium*, *Aspergillus*, *Alternoria*, *Cladosporium*, *Fusarium*, *Mucor* and *Stemphylium*.

A few of these species groups are particularly worth noting, as they are visible on seaweeds; they are all examples of ascomycetes or sac fungi. The parasitic *Phycomelaina laminariae* forms black patches on all kelp species, invading the tissue at the growth point on the stipe. On Sugar Kelp the infection manifests itself after a year of seaweed growth and the severity increases as the temperature of the water rises. As the infection develops it enhances the ability of saprophytic fungi to colonise the still-living kelp fronds. Interestingly, farmed kelp plants are twice as likely to be infected as wild specimens. Moreover, the presence of a dense layer of bacteria on the frond minimises the fungal attack, which suggests that the bacteria may be conferring some protection to the kelp. Other examples of common fungal infections include *Didymella fucicola* on fucoid algae and *D. magnei* on Dulse, where the disease causes the frond to be damaged and more susceptible to wave action. Some, such as *Lulworthia fucicola* and *L. kniepii*, are highly specific to a host, and these occur on Bladder Wrack and the calcareous red seaweed *Lithophyllum*, respectively. Not all are believed to be parasitic: *Mycophycias ascophylli* is common in the receptacles of Egg Wrack and Channel Wrack but appears to cause no harm and, as seen in Chapter 5, may actually help to enhance its host's resistance to environmental extremes.

The fact that fungi are present on and in living seaweeds is important for a couple of related reasons. First, fungal attacks increase the chances of a seaweed being damaged by wave action and ending up in the strandline. Second, if the seaweeds that are dominant in the strandline (wracks and kelps) are also the seaweeds most likely to have saprophytic fungi already well established on their surfaces, this means that, even if breakdown has not yet commenced, once the seaweed arrives in the strandline, decomposition will be off to a flying start. Two-thirds of marine fungi are in residence within the strandline and, along with bacteria, their spores will be dispersed among the newly arrived organic matter on the bodies of the flies and sandhoppers living there.

Sandhoppers

In midwinter, when the British Isles may be locked in frost and snow with insect life suspended in deep torpor, hibernation or diapause, invertebrates on the very edge of the land can be found not just surviving, but thriving. At first glance, the seaweed freshly deposited on the strandline in the preceding days may appear inert and lifeless, but a gentle nudge will soon cause tiny flies to come to the surface and fly a short distance away. For an even better demonstration of the population of small animals living on strandline material, try placing a bucketful of the drying, desiccating weeds on a white sheet and watching closely as creatures emerge. Three groups of animals usually dominate: sandhoppers, rove beetles and seaweed flies. These are not random species of insect or crustacean but a highly specialised, limited range of species forming a unique community. Every shore has such a community, and the component individuals can be found every day of the year, especially in the winter at times when seaweed is stranded in abundance.

Several species of sandhopper are strandline specialists, including *Orchestia gammarellus* and *Talitrus saltator*, although the latter is associated more with sandy than rocky shores. These species can grow up to 20mm in length, and they rapidly flex their abdomens to escape predators if disturbed, leaping a great distance – hence their common name of sandhopper. Huge numbers can be found at the bottom of decaying seaweed by day, burrowing down into the sediment (if present) when they are not feeding, and remaining where it is humid to avoid desiccation. Activity peaks soon after dark, when sandhoppers

A sandhopper: this small amphipod crustacean is found in huge numbers in all seashore strandlines throughout the year.

emerge from the base of the strandline to feed and disperse. Their behaviour is regulated by day–night rhythms as well as by tidal ones, and they are known to orientate themselves using the sun and moon.

Sandhoppers are usually the first scavengers to begin feeding on freshly deposited seaweed, devouring it and physically breaking the organic matter down into smaller particles for microorganisms to work on; they are consequently responsible for a huge release of mucilage (a gelatinous polysaccharide substance) from the seaweed, and mixed with their faeces this creates an exceptional amount of slime. The presence of the slime will be very noticeable to anyone lifting strandline debris with their hands, but the fact that it also coats the surface of the substrate on which the strandline has developed is perhaps less obvious. In the case of a sandy beach, individual sand grains will be bound loosely together by the slime, literally stabilising the shore. Some town councils have found to their cost that mechanical cleaning of beaches to remove strandlines (seaweed and sandhoppers) causes more problems than it solves. The financial asset of a pristine tourist beach is soon offset by the cost of removing tonnes of sand from the city centre when storms move destabilised beaches landwards. If there are no sandhoppers, there may soon be no beach.

Seaweed flies

Flies and rove beetles follow closely behind the sandhoppers to colonise newly stranded seaweed. Seaweed flies appear to have little in their nature to attract the attentions of even the most dedicated fly specialists, but these remarkable creatures work with the microbial populations to recycle dead organic matter faster than in any other system. The predominant genera of seaweed flies are *Thoracochaeta*, *Orygma*, *Coelopa* and *Fucellia*.

The genus *Thoracochaeta* is part of the large family Sphaeroceridae or lesser dung flies, which more typically live on organic matter such as cow pats. *Thoracochaeta* species, however, are specialist inhabitants of the strandline and can survive very salty conditions. When it is warm they form small clouds above the dead seaweed, but they are still very active even in midwinter. Little is known about their larval or maggot stages other than that they graze on microbes in humid, bacteria-rich areas. This is a worldwide genus that includes several weird and wonderful examples: one species lives in the antennal glands of land crabs, others in caves and on fungi.

ABOVE: A seaweed fly maggot (left) exposed on a piece of stranded sea lettuce; a rear view (right) of a seaweed fly maggot, showing the raised spiracles through which it breathes.

LEFT: A newly emerged adult seaweed fly climbing to the surface of the strandline. Its wings will not expand until the fly is well clear of the decayed seaweed, since released mucus and liquid would otherwise stick the wings together.

Another small fly, *Orygma luctuosum* (a member of the family Sepsidae), can have huge numbers of tiny thin maggots in the strandline, feeding on microorganisms.

The specialised association between maggots and microbes is unique in producing rapid and efficient decomposition in the short time between tide cycles, converting the food produced by photosynthesis on the rocky shore into soluble nutrients. The fly larvae that survive by microbial grazing require an astounding degree of specificity. One such species is the oily-looking Wrack Fly or Kelp Fly *Coelopa frigida*, one of the few flies that can escape from the water surface and not become trapped in the surface tension if the tide covers the strandline. Adults emerge from strandline kelp deposits in late autumn and can reach astronomical numbers, causing mass movements of flies searching for further breeding sites. Henry Egglishaw from Newcastle University studied Wrack Flies in the 1950s and 1960s and commented that they were one of the few fly species that specifically swarmed, showing mass movements a metre or two above the ground. Many millions can aggregate on a small area of strandline, and they can reach nuisance proportions when they also congregate on waste bins, signs and other structures.

Two swarm behaviour types have been identified in Wrack Flies: first, a mass flight when the tide returns or if the flies are disturbed; second, a mass migration at low tide when they move to a new breeding site. This is stimulated when the decomposition creates so much heat that the strandline becomes too hot, or when overcrowding reaches a critical level. Swarming is usually parallel to the beach initially, but if adults get blown off course then other stimuli come into play. Wrack Flies have a particular penchant for the organic compound trichloroethylene, which does not occur naturally (it is a liquid commonly used as an industrial solvent) but has a 'sweet' smell that could be similar to odours in the strandline. This was first reported in 1954 and investigated by the famed entomologist Harold Oldroyd when, in the autumn of that year, a huge swarm appeared across the Home Counties and reached London, probably owing to unusually persistent strong southwesterlies blowing the flies inland; the flies were a particular nuisance as they congregated around the shops of dry-cleaners.

The Wrack Fly is easy to rear and has become an alternative laboratory subject to the Fruit Fly *Drosophila melanogaster*. As a consequence, a great deal is known about its genetics (Leggett *et al.* 1996). Male body size varies considerably, so much so that in the past some individuals have been described as being members of different species. Their size is determined by a rather complex chromosomal inversion system, and larger males will overpower smaller ones to mate with females. However, the female is highly promiscuous, mating hundreds of times during her short life and storing individuals' sperm separately, even having the ability to pick and choose which she will use to fertilise her eggs.

Wrack Flies are often present in huge swarms in the strandline.

On the shore, female Wrack Flies lay many hundreds of eggs in fresh seaweed deposits; under ideal conditions the larvae pupate in a week to ten days. Most flies would be unable to tolerate salt in their diet, but Wrack Fly maggots are adapted to cope with and excrete it. Rotting seaweed provides warmth, and the compost provides a nourishing slime on which they can feed, using a hooked mandible to tear the seaweed and rake it into their mouths, which helps in the physical breakdown of strandline debris, opening it up for microbial access and more rapid recycling. The muscular, segmented maggot body surrounds a simple intestine and a pair of very large salivary glands. As the larvae wriggle through the seaweed they secrete digestive enzymes and spread putrefying bacteria, which in turn help to maintain their slimy environment. Maximum efficiency is achieved by the active hordes of maggots mixing the ingredients together. All fly maggots typically have a pair of spiracles at the rear end of the body so that they can continue to breathe while moving through the substrate (see p. 317); in the Wrack Fly maggot these hind spiracles are unusually prominent, with an elaborate fringing of hairs to keep fluids away.

Fucellia, another common genus of seaweed flies found around the strandline, are dark greyish flies that are close relatives of the pest Onion Maggot *Delia antiqua*, superficially resembling the Housefly *Musca domestica*. Although *Fucellia* can be abundant, they do not form the huge swarms commonly seen in the Wrack Fly. Individuals can be spotted across the shore, particularly in the vicinity of barnacles, around the edges of pools and among living wracks, looking for anything dead or decaying; fish and crab refuse are favourites. Small aggregations of flies around the shells of dead mussels are likely to be the species *Fucellia tergina*, especially during the summer. Although their larvae can be found elsewhere on the beach, the majority will be in the strandline. The larval diet resembles that of the Wrack Fly when the maggots first hatch, but as they age they develop an interest in fresh meat, favouring small insects and young sandhoppers.

In addition to the strandline-specialist genera, several species of a surprising array of other groups can also be found residing amongst the debris. At low tide many will also explore further down the shore – and like other opportunists they may feed there. The large predatory fly *Helcomyza ustulata* is more conspicuous than most as it darts forwards for a metre or more, searching for its next hiding place and for potential prey. Several flies of the genus *Aphrosylus* are associated with colonies of barnacles on rocks, stone walls and jetties, and at least one species has larvae that prey on living barnacles. The adults meanwhile eat the marine midge *Clunio marinus* and its larvae, which they find in the surface algal film between barnacles. *Aphrosylus* is a genus of the dolly flies or long-legged flies (Dolichopodidae), and four species occur in the British Isles.

Another common predatory fly exclusive to the shore is one of the dung flies, *Scathophaga litorea*. Compared with the bright golden-yellow terrestrial dung flies that are more familiar in other environments, this species is very drab, appearing dull grey and black like other intertidal flies. These carnivorous flies are as important as those that scavenge the microorganisms; by feeding on the latter they help the recycling process.

Beetles and more

Along with these predatory flies, some carnivorous beetles are also able to find a place to live on the strand. Occasional large carabid ground beetles are attracted down to the shore by the warmth in autumn and winter, but a whole host of specialist rove beetles (family Staphylinidae) inhabit the shore year-round; this is the only coleopteran family to have evolved such a close association with rocky shores. A few particularly large species of *Staphylinus* rove beetle (referred to as devil's coach horses) may be conspicuous when reacting

to disturbance by lifting their abdomen and head. Like *Staphylinus* beetles, *Aleochara obscurella* and *Cafius xantholoma* will be in search of young sandhoppers and smaller beetles. We have seen them attempting to take on maggots, but they are usually defeated by becoming embroiled in slime.

Arguably most significant is the range of much smaller (3–8mm long) rove beetles, which occur in huge numbers. Most are microbial grazers living their entire lives in the strandline. The *Omalium* rove beetles are 4mm long and differ from the usual rove beetle in that they have a broad flat body shape, a feature they share with the seaweed flies, though the wing cases still only cover the upper half of the abdomen, which is typical of the rove group.

Finally, as well as sandhoppers and insects, we should also mention the presence of mites, sometimes found in high densities. The genus *Halolaelaps* is typical of the seashore and some species live exclusively in the strandline, where they are predatory, possibly on springtails, and also suck blood from larger invertebrates.

ABOVE: Many specialist beetles, such as this *Omalium* rove beetle, are found only in the strandline. Inset: head of a predatory rove beetle.

The top strandline predators

With so much maggot protein available in the strandline, a variety of predators is inevitably attracted to feast at the top of the shore. The Rock Pipit is fairly common here, and while it flits between all the zones it has a particular liking for the strandline, where it can often be seen picking out the white maggots. The Turnstone also spends much of its time feeding in this area, especially when the debris is well rotted and not too deep (fresher seaweed will be in dense tangles and more difficult to turn over); using its short, stout beak, it is able to flick the material quickly over to feed on the sandhoppers and maggots beneath.

Other predators are likely to be transient or occasional visitors. The Pied Wagtail *Motacilla alba* is a regular sighting on the shore at any time of the year, arriving in pairs or singularly to feed on the insects. By contrast, great flocks of Starlings *Sturnus vulgaris* appear during the winter to feed in the strandline. Opportunistic scavengers in the crow family (Jackdaw *Coloeus monedula*, Carrion and Hooded Crow *Corvus corone* and *C. cornix*, and Raven *C. corax*) come to feed on the decomposer community or any suitable dead animal that has been

Turnstones feed in the strandline on invertebrates, typically sandhoppers.

Nature's giant compost heap

Pied Wagtails and Starlings commonly feed on insects and sandhoppers in the strandline. The wagtail (left) is an adult male; the Starling (above) is a juvenile moulting into its first-winter plumage.

washed up, whether it be a fish, a crab or a seal pup that failed to make it through the autumn gales. In one particular sheltered rocky shore in Pembrokeshire that backs on to a wood, it is not uncommon to see Noctule Bats *Nyctalus noctula* feeding at midday over the strandline, collecting seaweed flies. Every so often each bat will stop feeding and fly a short distance to a small pool of brackish water, dip down to drink, and then return to the seaweed flies. Maybe the flies are too salty, and the drinks of water are like sipping a beer in between bites of pork scratchings.

Shrews will also visit strandlines to take advantage of the concentration of insect prey, and voles sometimes use large banks of decaying seaweeds in winter to provide warm burrows. Even larger mammals such as deer will occasionally visit the strandline to supplement their diet by nibbling on washed-up seaweed. We have noticed this among Red Deer *Cervus elaphus* in the Scottish isles, but also with Roe Deer *Capreolus capreolus* along gravel-bordered estuaries in southern England.

One final example, a little outside the remit of this book, involves the world's largest land predator, the Polar Bear *Ursus maritimus*. Starving adults meet at Hudson Bay, Canada, to be first onto the sea ice to catch seals. While waiting on land for the ice to form, many search for other food that could sustain them. The shore in late autumn is a metre or so deep in decayed kelp, and the bears can be seen grazing on the choicest maggot infestations, the best available protein. The importance of the community in the strandline cannot be over-stressed.

A Red Deer feeding on stranded kelp seaweed.

The strandline in summary

Strandlines can be very untidy-looking places – areas of damage and decay, with varying amounts of human detritus. Yet this is a habitat that provides the main input of nutrients to intertidal communities. Organic matter arrives in varying quantities twice a day with every high tide to create the largest and most continuously maintained compost heaps in the world. Then, at phenomenal speed, it is broken down into smaller, simpler, soluble products by microbial action, and these products are redistributed by the tide. Decomposition is accelerated by specialist seaweed fly maggots, which in turn become food for beetles, birds and other carnivores. On exposed sandy shores, nutrients from strandline decay enter the sand or shingle, and provide suitable conditions for pioneering plants to establish themselves on the beach – which could, in due course, enable sand dunes to form. Members of terrestrial food webs benefit as land animals come to feed in winter on the nutrient bonanza. What the rocky-shore herbivores cannot do, the seaweed flies have achieved: they consume huge amounts of algal biomass and convert it into usable material to re-enter the ecosystem. Perhaps, despite the smell and the slime, strandlines do indeed deserve further study.

This driftwood has been stranded after drifting for many months at sea, where it was colonised by Goose Barnacles.

Challenges, threats and the future of rocky shores

chapter thirteen

Not only impressive, diverse and dynamic, rocky shores are also one of the few remaining habitats in 21st-century Britain and Ireland that might be described as natural. But herein lies a problem: what do we mean by 'natural', how does 'natural' compare with 'unnatural', and how do we tell the difference? If 'natural' means devoid of human interference, how do we tell if a shore meets this criterion? Is 'unnatural' necessarily bad?

Surprisingly, it is the humble limpet that has helped to answer some of these questions. But this is a story that involves not only limpets; Dogwhelks, barnacles, an oil spill, antifouling paint and hundreds of Dale Fort students also have a part to play.

Tales of Frenchman's Steps

In the 1980s, groups of A-level biology students would investigate the numbers and size range of a population of limpets on a rocky shore near Dale Fort Field Centre called Frenchman's Steps. Residential courses lasted six days, meaning that limpet data were collected by a new group of students virtually every week. For no particularly good reason, we decided to keep the results, just in case.

The students' research focused on how limpet numbers varied with height up the shore, reaching a maximum in the middle (see graph on p. 328). This makes perfect sense: at the top of the shore abiotic factors including dehydration and temperature stress make life unpleasant for marine animals; at the base, biotic factors predominate,

OPPOSITE PAGE:
A strandline replete with human refuse.

Frenchman's Steps study shore, looking towards Dale village.

and other species — barnacles and seaweeds on this particular shore — will compete with limpets for space. Optimal conditions exist in the middle shore between these two extremes, and here the limpets thrive.

We also measured limpet shell length and found that most were 10–15mm long (the largest limpet we have ever seen on a British shore was a little shy of 65mm); see graph opposite. We can make sense of this result if we assume the oldest limpets are the biggest. As limpets grow throughout their lives, this is not an unreasonable assumption (although it will not always be the case, as growth rates are very sensitive to local

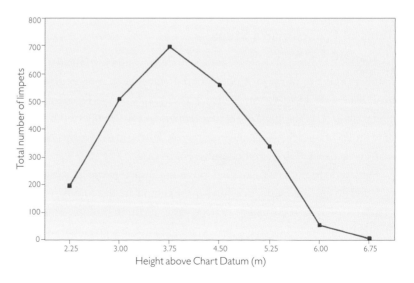

Total number of limpets at each height up the shore, Frenchman's Steps, 1985.

Challenges, threats and the future of rocky shores

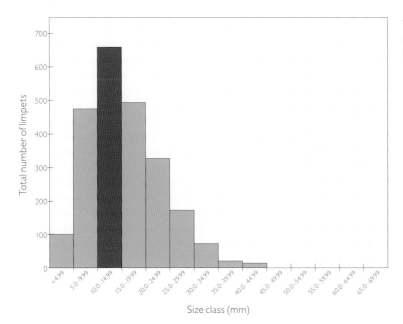

Total number of limpets in each size class, Frenchman's Steps, 1985.

food supply). We found fewer large (older) limpets, which makes sense, because they die of old age. It seemed strange that there were also fewer small (younger) ones, as this would suggest the population was doomed. However, small limpets are harder to see (easily mistaken for barnacles by students fairly new to rocky-shore fieldwork), so are likely to have been under-recorded. Young limpets also grow more quickly and may grow out of the smaller size classes quite rapidly.

In February 1996, the *Sea Empress* covered all our local shores in crude oil. Here was an event that was unequivocally 'unnatural' – but

The *Sea Empress* tanker wrecked off St Ann's Head, Pembrokeshire, February 1996.

A 'green flush' on the shore. Reduced numbers of grazing herbivores allow the fast-growing green algae to proliferate. This is a typical sign of a polluted shore – or indeed the welcome first signs of recovery.

to say anything meaningful about the effect of the oil it was necessary to know what was there before the spill. Fortunately, in terms of the limpet population on Frenchman's Steps, we most certainly did. All of the results from student groups going back 14 years were in a box file in our office; suddenly it became a very valuable data set.

Students investigating the limpet population after the spill were effectively also researching the effects of oil pollution on rocky shores and how quickly they recover. An obvious consequence, visible from some distance in the spring of that year, was the unusual 'green flush' on the shore. As the oil had reduced the numbers of grazing herbivores, namely topshells, periwinkles and limpets, relatively fast-growing green seaweeds had been able to thrive in their absence. Later still, brown algae increased in abundance but at a more leisurely pace.

In a memorable incident soon after the spill, one of the students nudged a limpet in the course of measuring it, and the limpet's shell fell off its back, a result of the animal being narcotised by hydrocarbons. Somehow this symbolised everything that was wrong about this new state of the shore, and the whole group was deeply affected. Within a year we were able to demonstrate that limpet numbers had returned to what might be considered 'normal', but the size data were skewed such that most limpets were now larger, at 15–20mm long. The obvious

A limpet (right) out of its shell (left). Shortly after the oil spill, when many limpets had already succumbed to hydrocarbon poisoning, this individual still clung to the rocks, but without the protection of its shell its days were numbered.

conclusion was that young limpets were particularly susceptible to oil pollution and had died.

Two years after the spill, limpet numbers had recovered spectacularly well, and to our delight the population size range had settled back to most limpets being 10–15mm again. It could be argued that the population had recovered in terms of numbers and size range to 'normal' within two years. The relatively quick recovery was, in part, due to the nature of the oil. Forties Blend is a light crude and at least one-third would be expected to evaporate from the sea during the first 24 hours after the spill. By moving with each high tide, the effect of this mobile oil in any one spot was somewhat ameliorated, although this was not the case on all the local shores.

As our limpet population data were so obviously useful, we decided to continue monitoring the shore to see what happened next. To negate any seasonal complications, data collection was limited to one month of the year. We got the heavy guns out in the form of some master's degree students from the University of Leuven to give us supremely reliable data sets; their visit, rather conveniently, was always in April. Now that the oil had gone and the limpet population appeared to be back to normal, we wished to investigate what sort of change, if any, would occur next under 'natural' conditions.

These magnificent students collected data for years, but in recent surveys Dale Fort teaching staff have taken over the reins (measuring limpets is riveting, but eventually even master's students need a break). From 1998 onwards the limpet population seemed settled in its post-pollution state, which appeared to be identical to the pre-pollution state, the state we had come to assume was 'normal'. Then, in 2010, something very peculiar happened (see graph below). Some

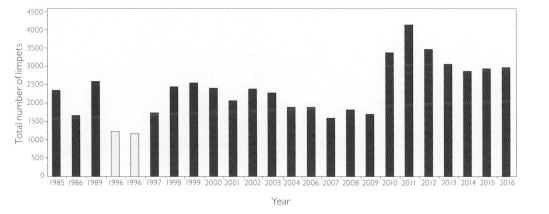

Total number of limpets on Frenchman's Steps, 1985–2016. (Two data sets have been included for 1996 to show how similar – and hence reliable – two A-level groups' results were, a fortnight apart).

conscientious members of staff had collected the data, so we had no reason to doubt their validity, but the results were off the chart. We had never seen so many limpets on Frenchman's Steps, ever. We did not comment on it at the time but suspected that these meticulous fellows had diligently measured and counted all the barnacles on the shore as well. So much so that the following spring one of us (JA-T) collected the data with colleagues, just to check that everything was as it should be. That year there were even more limpets again. Indeed, 2011 was the year in which the record number of limpets was found on this shore. In the years between 2011 and 2016, numbers appeared to settle back down again, but only to the new level established in 2010. A new 'normal' has emerged.

Looking at shore totals (see graph on p. 331), or mean numbers at each height (see graph below), we seem to have a pre-pollution 'normal' of approximately 1,500–2,500 individuals between 1985 and 1989, and a pollution state of approximately 1,200 limpets in 1996. Between 1997 and 2009, numbers were back to the pre-pollution norm of 1,500–2,500 again. Finally, we have a post-2010 population of 2,800–4,100 individuals. What on earth is going on?

To answer this, we have to bring in some other shore dwellers and anecdotal observations from the shores over the decades. In 1982, there were virtually no Dogwhelks on the shores between the Field Centre and nearby Dale village; a search on local shores might yield two or three animals in an area of 1,500 square metres. Dogwhelks are

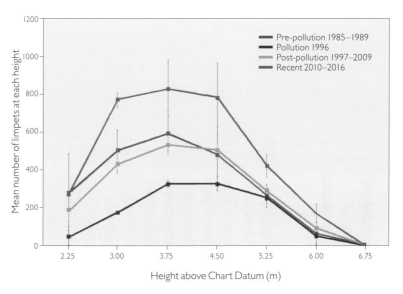

Mean number of limpets at each height up the shore for four shore states, with 95 per cent confidence limits.

fairly common rocky-shore animals in the British Isles, as we know, but at the time their numbers were severely depleted by tributyltin (TBT), a chemical leached from antifouling paint, which was widely used to prevent fouling organisms (barnacles, seaweeds and so on) from growing on the hulls of pleasure boats and larger commercial vessels.

In the 1980s unusual changes in the sexual characteristics of Dogwhelks were being noticed in estuaries and other areas where small boats, painted with TBT antifouling paint, were concentrated. Female Dogwhelks were growing a non-functional male reproductive organ, preventing successful reproduction – a condition termed 'imposex'. As a result, over about 20 years, many local snail populations became extinct. TBT paint was so toxic that imposex could be initiated in newly hatched snails at concentrations of only 2 nanograms of tin per litre of seawater (less than a teaspoon-full in an Olympic swimming pool). Fortunately, the International Maritime Organization banned TBT antifouling paints from small craft (less than 25m long) in 1985 and from all craft in 2003, and all traces of the paint had to be gone from hulls by 2008.

After the ban, Dogwhelk populations began to recover on the shores around Britain and Ireland. On those near Dale Fort Field Centre this recovery was so successful that local shores are now host to unsustainably high numbers. Dogwhelks typically eat barnacles, especially the acorn barnacle *Semibalanus balanoides*, which is abundant at Frenchman's Steps. One theory for the spectacularly high limpet

Dogwhelks on Frenchman's Steps in recent years have shown an unsustainably high population density.

densities in 2010 and 2011 is that the now-abundant Dogwhelks may have eaten so many barnacles that more space was available for limpets. After 2011, the number of limpets started to decrease again, and since Dogwhelks are known to eat young limpets, which were present in unusually large numbers on the shore in 2011, it is possible that, having had a significant impact on the barnacle population, Dogwhelks began eating small, thinner-shelled limpets as well. Since 2013 the total population density of limpets seems to have stabilised, albeit at a higher than 'normal' level.

Hopefully these decades of observations have highlighted the difficulties in establishing whether a particular shore state is affected by human activity or not. The apparently 'normal' limpet population state before the oil spill was probably an expression of TBT effects, in that the lack of Dogwhelks in the 1980s allowed a higher than previous population density of barnacles to exist, with the result that the limpet numbers were suppressed. As Dogwhelk numbers recovered after the TBT ban, barnacle densities may have been depressed, allowing limpet numbers to increase again. Thus, the recent data may not represent an unnaturally high limpet density at all, but a recovery to pre-TBT levels (Archer-Thomson 1999, 2016).

These examples at Dale Fort are perfect illustrations of chronic and acute pollution. An oil spill such as that from the *Sea Empress* is an acute type of pollution, which is caused by a single event and then disappears to allow recovery. An exposed rocky shore, with lots of wave action and water movement to break up the oil, might recover from a spill within a couple of years, a sheltered shore within four to six years, while an extremely sheltered place such as a saltmarsh might take 20 years or more. With chronic pollution such as that from TBT, the effect, as experienced by the unfortunate Dogwhelks, is often low-level but goes on for as long as the poison is present in the environment.

Other pollutants

Of course oil is not the only pollutant that threatens marine ecosystems. We as a species throw all sorts of chemicals into rivers and estuaries, and ultimately all watercourses lead to the sea. Agricultural runoff in the form of nitrates and phosphates fertilises phytoplankton communities just as well as grassland ones, and the effects are similar. Those species that can use the extra nutrients do well at the expense of those that cannot, and community composition is altered accordingly.

As the rocky shore is both a major contributor to and beneficiary from the plankton, anything that affects the plankton community will also affect the shore, not to mention the knock-on effects throughout all marine food webs. The *Ulva* species (sea lettuce and gutweed) proliferate when exposed to agricultural runoff, while soil washed off from the land can smother subtidal and shore communities to the particular detriment of filter feeders. In addition, these soil particles attract other pollutants and so can serve to concentrate toxins in 'sinks' where sediment accumulates.

There are many recognised definitions of pollution, but the most poignant one we have ever come across is that it is 'an environmental insult'. Humans insult marine and coastal systems a lot. There are parasites, bacteria, viruses and various hormones from human sewage (although sewage treatment has improved dramatically over recent decades). There are also organic compounds including carbon-based chemicals from dairies, slaughterhouses, sewage farms, breweries and tanneries, all of which are food for microorganisms whose presence can affect inshore waters and shore inhabitants. Industrial contributions are many and varied; they include acids, alkalis, mineral fibres, heavy metals, radioactive substances, and even inert suspensions from mines and quarries that smother rocky-shore communities and clog fish gills.

Thermal discharges include power-station outlets that release coolant water above the ambient temperature. A local example in the Milford Haven Waterway releases water at 8°C above background, and rocky-shore surveys near the outlet pipe show impoverished communities and

Warm water is discharged from Pembroke Power Station, and the foam is a result of turbulence aerating a mixture of dead plankton and other organic detritus.

the presence of non-native species more tolerant of warmer water – for example the Red Ripple Bryozoan *Watersipora subatra*, whose origin appears to be uncertain at present. This all happens within a Special Area of Conservation (SAC). Another invader, the Leathery Sea-squirt *Styela clava*, this time from the northwest Pacific, was first recorded in Plymouth Sound in 1953, probably having hitched a ride on the hull of a warship returning from the Korean War for maintenance in the naval dockyard in Plymouth. It does particularly well in areas where there is (artificially) warm water.

Another pollutant that has received a lot of attention in recent years is plastic. According to Greenpeace, in the spring 2017 edition of their *Connect* magazine, approximately 12 million tonnes of plastic waste enters the oceans each year, equivalent to 35,000 tonnes a day, or a rubbish truck dumping its contents into the sea every minute. This annual input joins the estimated 150 million tonnes that is already there. By 2050 there could be more plastic, by weight, than fish in the marine system (especially if overfishing continues apace). Plastic litter is universal, occurring on beaches and rocky shores everywhere, in the open ocean, in the water column and even as deep as the ocean trenches. Turtles eat it, mistaking it for their usual jellyfish fare, seals, whales, dolphins and porpoises get tangled up in it, and seabirds get it caught around their legs. A Marine Conservation Society factsheet from the 1980s claimed that a million seabirds and 100,000 marine mammals worldwide were killed every year by getting tangled up in or ingesting plastic.

A Gannet with fishing twine bound for incorporation into the nest material. Many young birds on the Grassholm gannetry have to be freed by the RSPB each autumn as their feet are entangled in plastic debris.

These are all well-documented threats to the marine environment, but a new issue has recently become more apparent. There are minute particles of plastic in the oceans now, derived from pre-production microplastic resin pellets or 'nurdles', microbeads from cosmetics, and breakdown products of plastic litter. In marine food chains, who-eats-whom is generally governed by size; organisms tend to eat anything smaller than themselves. There are plastic particles small enough now to be ingested by plankton, where they give no nutritional benefit, can be directly toxic and may block digestive pathways. Once ingested, this material can pass up the food chain, occasionally killing as it goes, and eventually it may end up in fish consumed by humans. Nurdles are present in strandlines all around the world. Happily, attention is finally being focused on this environmental issue such that a ban on products containing microbeads is being rolled out in many countries.

Nurdles are present in most strandlines in the world in this 'plastic age'.

Now there is a new twist to the plastic tale. Matthew Savoca of the University of California has found that when plastics are exposed to seawater they emit a volatile chemical called dimethyl sulphide (DMS). This same chemical is emitted by phytoplankton, and seabirds – Fulmars, for example – use this gas as a sign of where food is likely to be. Ironically it seems that seabirds are now actively attracted to the very plastics that they would do well to avoid. Savoca's (2016) study indicates that seabirds that use DMS to locate food are six times more likely to ingest plastic waste.

A Fulmar's acute sense of smell allows it to detect dimethyl sulphide, which usually leads it to food. But in recent years this has also led the birds to plastic, with unfortunate consequences.

Habitat destruction and mitigation

As humans continue to exploit the coast for industrial, commercial and leisure activities there will inevitably be effects on rocky-shore communities. These might include direct destruction as natural coast is replaced by hotels, industrial structures, marinas, harbours, jetties, roads, rail links and so on, or indirect effects such as increased sedimentation, decreased water quality and increased turbulence affecting the shore and adjacent habitats. However, there is a movement now to add to artificial coastal structures with so-called eco-engineering to enhance biodiversity. Examples include drilling artificial rock pools and pits into breakwaters, at the time of construction or retrospectively, or incorporating parallel longitudinal depressions along the length of coastal defence blocks to increase their structural complexity, thereby providing little crevices for barnacles, mussels, sponges, seaweeds, periwinkles and other organisms. A very readable summary of such initiatives, with some actual examples, is provided by Firth *et al.* (2012).

If structures are engineered with features that improve local biodiversity, it could be the case that 'artificial' need not be bad or even worse than 'natural'. After all, coppiced woodland and grazed meadows are heavily shaped by human activity, but can be delightful habitats of high conservation value.

Climate change

Almost everyone now subscribes to the view that human activity is affecting global climate. Our global annual emissions of carbon dioxide, approximately 49 billion tonnes, alter the planetary heat budget and, as a result, disrupt weather patterns and ocean currents. There is also concern about melting polar ice caps and rising sea levels. So far, estimates suggest that the oceans have absorbed 40–50 per cent of our post-industrial-age carbon output. This has slowed the predicted rate of climate change but cannot be expected to continue indefinitely. There is a price to pay for this service: when carbon dioxide is absorbed by seawater it causes a decrease in the pH, in other words the sea becomes more acidic (there has been a 30 per cent decrease in pH since 1800).

How do these global events impact rocky shores around the British Isles? Plankton can be affected, for one. The larvae of some sea urchins, such as the Shore Sea Urchin, have calcareous spicules for protection against predators and to increase their buoyancy by increasing their surface area, but these spicules will not form in acidified water – and acidification has been shown to directly affect survival. Many rocky-shore mollusc larvae, and indeed shelled organisms in general, are in trouble because acidified water causes their shells to dissolve. Anything that affects larval recruitment to the shore can have profound effects on intertidal communities. Ocean acidification may also have an effect on adult molluscs: the Edible Periwinkle responds to the presence of predatory crabs by producing a thicker shell, but this reaction is reduced in seawater of a lower pH.

In short, any organism that relies on the production of calcium carbonate structures for shells, coral skeletons, planktonic spicules and so on will be severely stressed by the new pH regime. When we were learning about photosynthesis at school in the 1960s, we were informed that atmospheric carbon dioxide concentration was 300 parts per million. Nowadays the figure is over 400; if it passes 450 then shell-dissolving conditions will be a permanent feature in polar seas and reef erosion will exceed reef building. These severe conditions will almost certainly have ramifications for the rocky shores of Britain and Ireland.

Sea temperatures around the British Isles have warmed by 1°C since the 1980s, and warm-water species may be expected to expand their range northwards as a result. The Black-footed Limpet, which at present is restricted to southwest Britain, is forecast to replace the Common Limpet as the climate warms. Bladder Wrack is also predicted to suffer directly from raised temperatures, and its abundance affects barnacle, mussel and limpet populations as they all compete for the same real estate on the middle shore.

The Toothed Topshell appears already to have extended its range northwards, although it is important to view this in context. We have studied the population dynamics of Toothed Topshells on a shore near to Frenchman's Steps (and of similar aspect and exposure) since the late 1980s (see graph on p. 340). Fortunately we took our first look at these animals in 1987, when the population seemed perfectly healthy. The method we used, again with A-level students, was broadly similar to that for the limpets; we counted and measured individuals, this time in a fixed sample area, at various heights up the shore from 1m above Chart Datum until we ran out of topshells. The maximum height for these animals was 6m. In 1988, for reasons unknown, the population had crashed from an apparently 'normal' 300 individuals within the study area to fewer than 100. This decline, it turns out, was noted in

Rocky Shores

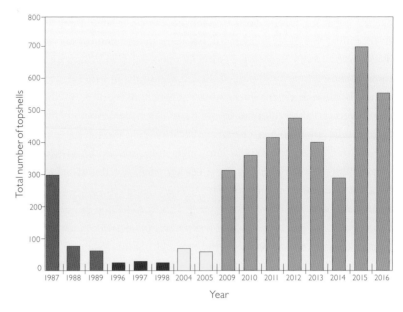

Total number of Toothed Topshells in the study area near Dale Fort. Different coloured bars denote gaps in the time-series data.

southwest Britain as a whole. The situation worsened in 1989. It wasn't until the *Sea Empress* year of 1996 that we looked at this population again. Unsurprisingly, numbers had fallen again, to half the already impoverished 1989 total. We took further samples throughout the late 1990s and early 2000s as time and student requirements allowed, but in 2009 noticed that there were more Toothed Topshells on the shore than we had seen for some time. Continuous data collection started in July of that year and has continued annually to the present.

In general, with a few hiccups along the way, numbers have increased in line with the expectation that this species will do well as the climate warms, but there is an important point that needs to be made here. Had we just looked at this population in 1987 in isolation (when all seemed well) and then after the oil spill in 1996, we would have blamed the oil for the massive population crash. We would have missed the previous downturn in the animal's fortunes that had happened, for reasons unknown, in 1988. It is only by looking at populations and communities for long – preferably continuous – periods that useful, more accurate, assessments of ecosystem change can be made. Long-term data sets are worth their weight in gold.

It is this premise that has been one of the driving forces behind the MarClim (Marine Biodiversity and Climate Change) project, a multi-partner-funded initiative that was created in 2000 to assess and predict the influence of climate change, using rocky-shore creatures as indicators. It uses historical data from the 1950s onwards as well as contemporary survey work to look for changes in abundance and distribution and to relate findings to recent rapid climate warming. The project encompasses the whole of Britain and Ireland, which are uniquely placed for change to be assessed as they lie in a transition zone between northern 'Boreal' waters and southern 'Lusitanian' ones, where many species are at the edges of their geographic range. Such species are more likely to be sensitive to temperature change.

Key findings have confirmed that typically southern, warm-water species including the Toothed Topshell (left), the Purple Topshell, the barnacles *Chthamalus stellatus* and *Perforatus perforatus* and Brown Tuning Fork Weed *Bifurcaria bifurcata* have all extended their range northwards by up to 150km since the 1950s. Eastward range extensions along the English Channel have also been observed in Toothed Topshells, Purple Topshells, China Limpets, Black-footed Limpets, Small Periwinkles, Strawberry Anemone *Actinia fragacea* and *P. perforatus*.

As might be expected, northern species such as the seaweed Dabberlocks and Tortoiseshell Limpet have shown contractions in their range at the southern edge. But there are more subtle effects as well. We mentioned in Chapter 2 how the barnacle *Chthamalus stellatus* (a southern species) competes with *Semibalanus balanoides* (a northern species) for space on the shore. It seems the balance of this competitive exclusion experiment is being tipped in the stellate barnacle's favour as warmer springs are adversely affecting the acorn barnacle's peak settlement phase from the plankton at this time of year.

As well as demonstrating the reality of climate change in plankton and rocky-shore communities, MarClim's findings also have repercussions for Marine Protected Areas (MPAs), which we will return to shortly. More details can be found on the Marine Biological Association's (MBA) website at www.mba.ac.uk.

Invasive species

Another consequence of warmer seas, coupled with global maritime shipping movements, is the introduction of alien species, and this is an extra thread to MarClim's work. More than 36 non-native seaweeds and marine animals (the number increases regularly) may now be found in ports and marinas, on boat hulls, on fishing gear or aquaculture equipment, and on natural shores. They include three species of brown seaweed (one of which is the familiar Wireweed mentioned in Chapter 1), six species of red, one species of green

Wireweed flourishing in a rock pool.

(Green Sea Fingers *Codium fragile* ssp. *fragile*), a sponge, an anemone, a comb jelly, a tubeworm, five molluscs, four sea-mats, two barnacles, two crabs, a shrimp and seven sea-squirts. In other words, these alien species are from many of the groups we have described in this book. Space does not allow a detailed look at all these new residents, but two are discussed a little further below.

Wireweed is native to the Pacific coasts of Japan, Russia and China. It first appeared on the Isle of Wight in 1973 and has since been recorded from the Essex coast, through the English Channel, up the west coast of Wales and rather more patchily all the way up to Orkney. Northern and western Ireland have also been invaded. It is thought to have been imported with Japanese Oysters *Crassostrea gigas*. This seaweed can grow up to 10cm a day and reproduces sexually from floating fragments that drift on ocean currents, which would help to explain its impressive colonising abilities. It favours sheltered areas and thrives in rock pools, where it can dominate by out-competing other rock-pool brown seaweeds such as Sea Oak *Halidrys siliquosa* and shading out understorey species. On the plus side, the seaweed does provide shelter for fish larvae and crustaceans, and cuttlefish use its fronds as a support for their eggs.

Slipper Limpets *Crepidula fornicata* originated from the east coast of Canada and the Americas down to the Caribbean. They reached Essex as early as 1887 as an accidental import alongside American Oysters *Crassostrea virginica*, although other routes such as in ballast water or on the hulls of vessels have been proposed. By the 1970s they were the dominant seabed creatures in the Solent, and when Henry VIII's warship the *Mary Rose* was raised from just outside Portsmouth Harbour in 1982 it was found to be covered in Slipper Limpets. The BBC's *Blue Peter* report at the time suggested that these delicacies were probably eaten by the Tudors!

Currently, Slipper Limpets are found throughout the entire southern coast of England, and personal observations can confirm that they are thriving on lower shores in southwest Wales. They have also been found up the coast in Cardigan Bay, and even in Scotland and the Hebrides. Although unwelcome, Slipper Limpets are rather interesting in that they live in a chain in which the lower individuals are female, the smaller individuals at the top are male, and those in the middle are changing sex. Males remain as such for up to six years and are 'encouraged' to do so by hormones released from the female at the base of the stack. Sex change takes about 60 days and only occurs in the lowest male – and presumably when he has become a she, the next animal will change sex, and so on. We have witnessed Slipper Limpet stacks occurring on both the sides and the top of a Common Whelk *Buccinum undatum* shell, and the poor host must have consumed a lot of extra energy carting these unwelcome guests about. Even so, these limpets are not immune to predation: Common Starfish and Common Shore Crabs will eat them, though they seem to prefer Common Mussels.

As well as disrupting rocky-shore communities, there is also an economic cost associated with these 'space invaders'. Oyster and mussel farm workers have to clean the fouling organisms off their stock before they can transport or sell them, and as these farmed bivalves are filter feeders, the Slipper Limpets are also competing directly for food.

Slipper Limpets covering the shell of a Common Whelk. Note a couple of *Spirobranchus* tubeworms getting in on the act as well.

Conservation

A common denominator and signature effect of human interference in natural systems is that of reduced species diversity. Conservation, which encompasses preservation and appropriate management, attempts to reverse these ill effects and enhance biodiversity. Few people would object to these laudable aims, unless perhaps they interfere with their livelihood and short-term economic gain. A Royal Commission set up in 2004 to investigate the state of the sea around the UK came up with some staggering findings concerning the management of our Exclusive Economic Zone (EEZ), the sea zone prescribed by the UN Convention on the Law of the Sea in which a state has special rights over marine resources, conventionally out to 200 nautical miles (but with complications where EEZ boundaries overlap). The commission concluded that for the oceans in the EEZ to recover from overexploitation, 30 per cent should be protected as non-exploitation zones. In other words, there should be areas around the UK where no one is allowed to take anything out of the ocean or put anything in.

To meet our commitments under the Convention on Biodiversity (Rio Earth Summit 1992) and to achieve 'good environmental status' across Europe's seas by 2020 (EU Marine Strategy Framework Directive), the UK's Marine and Coastal Access Act (2009) allowed for the creation of Marine Protected Areas (MPAs). As a result, the UK committed to create an ecologically coherent network of MPAs by 2012, and that they should be 'well managed' by 2016. To put it bluntly, we have failed miserably in this endeavour, despite valiant efforts from the Marine Conservation Society, local Wildlife Trusts and a host of other organisations and individuals.

According to the Joint Nature Conservation Committee (JNCC) website, updated in March 2018, 24 per cent of UK waters are currently within MPAs: there are 105 Special Areas of Conservation (SACs) and 107 Special Protection Areas (SPAs) with marine components, 56 Marine Conservation Zones (MCZs) and 30 Nature Conservation Marine Protected Areas (NCMPAs). There are other contributing designations (SSSIs and Ramsar sites) that will also help. But these are not non-exploitation zones. At present, there are only three such zones, namely Lundy Island in the Bristol Channel, Flamborough Head in Yorkshire and Lamlash Bay on the Isle of Arran. Together they occupy 0.01 per cent of UK waters. A network of no-take reserves has been proposed, but weak, or non-existent, governmental will and staunch opposition from commercial fishermen amongst others has stalled the process time and again. Some progress is being made, though, and the recent approach, in conservation as a whole, to concentrating on protecting larger areas is a giant step in the right direction. MarClim's findings add another layer of complexity to the problem, however. Any network of reserves that is envisaged, which must include replicates of each type of representative habitat, will have to allow for boundary movements as a result of climate change and associated species migrations.

Anyone interested in the details of marine (and terrestrial) conservation developments will be able to find staggering amounts of information online with the appropriate search-engine input. However, be prepared for a terminological jungle, which we have tried to avoid here: OSPAR, WSSD, Ramsar, EC Wild Birds Directive, EC Habitats Directive, SPAs, SACs, EMSs, Natura 2000, MSFD, WFD, SSSIs, NNRs, MCZs, MNRs, MPAs and many more. It is a salutary point that the most powerful legislation on conservation for terrestrial and marine habitats in recent years has come from Europe. Even several years after the referendum, Brexit sends shivers of fear down the spines of most people involved in protecting our natural heritage from destruction. The optimistic view is that we should be able to use our independence to create stronger protection for our beleaguered national biodiversity and not just use it as an excuse to kowtow to economic imperatives. Environmentally sound development can of course be cost-effective in the long run and protect biodiversity at the same time, as demonstrated in two excellent books by Tony Juniper, former director of Friends of the Earth, namely *What Has Nature Ever Done for Us?* (2013) and *What Nature Does for Britain* (2015).

Skomer MCZ, Pembrokeshire.

Fortunately, unlike their terrestrial counterparts, all that MPAs require by way of appropriate management is that we leave them alone, monitor them and protect them from harm. This is the case for rocky shores too. Look at them, photograph them, study them, even love them, but definitely protect them from all the negative influences mentioned above. Rocky shores by their very nature are resilient habitats. From taking student groups to the same shore for over 30 years it is abundantly clear to us that all of the trampling we do is utterly insignificant in comparison to the damage inflicted by one good storm. Even so, a strict code of conduct applies when we study a shore. Any boulders that are moved are replaced carefully, the right way up. Animals are put back where they are found. Seaweeds and lichens are identified *in situ* as far as possible. Collection is avoided if feasible, and absolutely prohibited for rarities. We hope that encouraging students, and indeed all visitors, to treat the shore with respect will translate into a more considerate approach to the wider natural environment. This would be a good thing.

Long-term prognosis

With the never-ending deluge of bad news concerning the state of our oceans and life on land, it would be all too easy to get profoundly depressed and adopt a doom-and-gloom attitude. Many young people studying for environmental degrees seem to suffer from this malaise. It is essential that we do not give up hope; a younger generation with no hope for the planet's future would be disastrous in many ways. Fortunately there is also good news, and progress on protecting the marine environment is being made. The spring 2017 edition of *BBC Wildlife* magazine championed the uplifting website www.oceanoptimism.org and the associated Twitter campaign that has a social reach of over 60 million: see #OceanOptimism and related campaigns #EarthOptimism and #ConservationOptimism. Other signs of increasing awareness can be seen in initiatives such as Sky TV's Ocean Rescue campaign, and in the 'Blue Planet effect' that was triggered by the BBC television series *Blue Planet 2* in late 2017.

So, although there are serious causes for concern about the state of the natural world including the marine environment, there is still some room for optimism. Visitors to the coast are often amazed at the diversity of life between the tides, and as a result come to care more about its welfare. Shores are promoted as places to relax and unwind, to discover the beauty and excitement of rock pooling, and not least as a location where people can connect with nature and share this connection with their children. More 'seashore safaris' than ever before are being run by charitable organisations including the Wildlife Trusts, the National Trust, the RSPB and the Field Studies Council, and they are invariably over-subscribed. The same is true of beach litter-picking events, originally championed by the Marine Conservation Society but now more widely organised. Beautifully illustrated books about seashores abound. Environmental education continues to evolve, and its powerful influence, both 'in the field' and, inevitably, online, is felt more than ever.

Ultimately, in about seven and a half billion years, when our sun goes supernova, all rocky shores are doomed – though admittedly this may be the least of our concerns at the time. Before that happens, continental drift (still rather a racy new theory when we were at university but now an established part of modern geology) will continue to move continents about the planet's surface and remodel coastlines in interesting new ways, the earth's climate will change as it always has done, affecting all terrestrial and marine ecosystems accordingly, and sea levels will rise and fall with ice ages and warmer interludes.

Meanwhile, we live in a geological epoch called, unofficially at present, the Anthropocene, named after the dominant factor affecting the earth's ecosystems: us. This could all change in an instant if a super-volcano erupts or a sizeable asteroid hits, but for now we as a species are presiding over the sixth mass extinction event in the planet's history – and this time it's our fault. All we humans can do, for the foreseeable future, is look after our life-support systems that make up the biosphere as best we can, rocky shores included, or we may simply not be around to benefit in the long term.

References and further reading

Åberg, P. 1992. A demographic study of two populations of the seaweed *Ascophyllum nodosum*. *Ecology* 73: 1473–1487.

Admiralty Tide Tables, Volume 1: European Waters. There is an alternative version, namely Admiralty Publication NP120, *Admiralty Manual of Tides* (1961). UK Hydrographic Office, Taunton.

Aldersey-Williams, H. 2016. *Tide: the Science and Lore of the Greatest Force on Earth*. Viking/Penguin, London.

Archer-Thomson, J. 1999. The *Sea Empress* incident and the limpets of Frenchman's Steps. *Field Studies* 9: 531–546.

Archer-Thomson, J. 2007. *The Chronicles of Larry, Volumes 1 and 2*. www.lulu.com.

Archer-Thomson, J. 2016. The *Sea Empress* incident and the limpets of Frenchman's Steps, twenty years on. *Field Studies*. http://fsj.field-studies-council.org (accessed July 2018).

Ballantine, W. J. 1961. A biologically-defined exposure scale for the comparative description of rocky shores. *Field Studies* 1 (3): 1–19.

Barber, J. H, Lu, D. & Pugno, N. M. 2015. Extreme strength observed in limpet teeth. *Journal of the Royal Society, Interface* 12 (105): 20141326.

Barnes, R. S. K. & Hughes, R. N. 1982. *An Introduction to Marine Ecology*. Blackwell, Oxford.

Barrett, J. & Yonge, C. M. 1958. *Collins Pocket Guide to the Seashore*. Collins, London.

Bengtson, S., Sallstedt, T., Belivanova, V. & Whitehouse, M. 2017. Three-dimensional preservation of cellular and subcellular structures suggests 1.6 billion-year-old crown-group red algae. *PLoS Biology* 15(3): e2000735. doi:10.1371/journal.pbio.2000735.

Benkendorff, K., Rudd, D., Nongmaithem, B. D. et al. 2015. Are the traditional medical uses of Muricidae molluscs substantiated by their pharmacological properties and bioactive compounds? *Marine Drugs* 13: 5237–5275.

Binns, R. 1969. The physiology of the antennal gland of *Carcinus maenas* (L.) [Parts I–V]. *Journal of Experimental Biology* 51: 1–45.

Branch, G. M. 1981. The biology of limpets: physical factors, energy flow and ecological interactions. *Oceanography and Marine Biology: an Annual Review* 19: 235–280.

Breidbach, O. 2005. *Art Forms from the Ocean: the Radiolarian Prints of Ernst Haeckel*. Prestel, London.

British Phycological Society & British Museum (Natural History). 1977–. *Seaweeds of the British Isles*. British Museum (Natural History), London.

Brodie, J., Maggs, C. A. & John, D. M. (eds). 2007. *The Green Seaweeds of Britain and Ireland*. British Phycological Society, London.

Bunker, F. StP. D., Maggs, C. A., Brodie, J. A. & Bunker, A. R. 2017. *Seaweeds of Britain and Ireland*, 2nd edition. Wild Nature Press, Plymouth.

Caldo, R., Dionísio, G. & Dinis, M. T. 2007. Decapod crustaceans associated with the snakelock anemone *Anemonia sulcata*. Living there or just passing by? *Scientia Marina* 71: 287–292.

Campbell, A. & Nichols, J. 1994. *Hamlyn Guide to the Seashore and Shallow Seas of Britain and Europe*. Hamlyn, London.

Caramujo, M.-J., de Carvalho, C. C., Silva, S. J. & Carman, K. R. 2012. Dietary carotenoids regulate astaxanthin content of copepods and modulate their susceptibility to UV light and copper toxicity. *Marine Drugs* 10: 998–1018.

Connell, J. H. 1961. The influence of interspecific competition and other factors on the distribution of the barnacle *Chthamalus stellatus*. *Ecology* 42: 710–723.

Costa, D. P., Marques, S. S., Baptista, T. M. *et al.* 2016. Preliminary study on the reproduction of the beadlet anemone *Actinia equina* (Linnaeus, 1758). Conference abstract, International Meeting on Marine Research 2016. *Frontiers in Marine Science* 3. doi: 10.3389/conf.FMARS.2016.04.00041.

Crane, N. 2016. *The Making of the British Landscape.* Weidenfeld & Nicolson, London.

Cremona, J. 1988. *A Field Atlas of the Seashore.* Cambridge University Press, Cambridge.

Cremona, J. 2014. *Seashores: an Ecological Guide.* Crowood Press, Marlborough.

Crothers, J. H. 1985. Dog-whelks: an introduction to the biology of *Nucella lapillus* (L.). *Field Studies* 6: 291–360.

Crothers, J. H. 2012. *Snails on Rocky Sea Shores.* Naturalists' Handbook 30. Pelagic Publishing, Exeter.

Crump, R. G., Morley, H. S. & Williams, A. D. 1998. West Angle Bay, a case study. Littoral monitoring of permanent quadrats before and after the *Sea Empress* oil spill. *Field Studies* 9: 497–511.

Dawkins, R. & Wong, Y. 2016. *The Ancestor's Tale: a Pilgrimage to the Dawn of Life.* Weidenfeld & Nicolson, London.

Dillehay, T., Ramírez, C., Pino, M. *et al.* 2008. Monte Verde: seaweed, food, medicine, and the peopling of South America. *Science* 320: 784–786.

Dobson, F. S. 2011. *Lichens: an Illustrated Guide to the British and Irish Species*, 6th edition. Richmond Publishing, Slough.

Douek, J., Barki, Y., Gateño, D. & Rinkevich, B. 2002. Possible cryptic speciation within the sea anemone *Actinia equina* complex detected by AFLP markers. *Zoological Journal of the Linnean Society* 136: 315–320. doi:10.1046/j.1096-3642.2002.00034.x.

Dring, M. J. 1986. *The Biology of Marine Plants.* Edward Arnold, London.

Dyer, B. D. 2003. *A Field Guide to Bacteria.* Cornell University Press, New York.

Egglishaw, H. J. 1960. Studies on the family Coelopidae (Diptera). *Transactions of the Royal Entomological Society of London* 112: 109–140.

Emson, R. H. & Crump, R. G. 1979. Description of a new species of *Asterina* (Asteroidea), with an account of its ecology. *Journal of the Marine Biological Association of the United Kingdom* 59: 77–94.

European Space Agency 2012. Lichen can survive in space: space station research sheds light on origin of life; potential for better sunscreens. *ScienceDaily.* www.sciencedaily.com/releases/2012/06/120623145623.htm (accessed July 2018).

Fincham, A. A. 1984. *Basic Marine Biology.* British Museum (Natural History), London.

Firth, L. B., Thompson, R. C. & Hawkins, S. J. 2012. *Eco-Engineering of Artificial Coastal Structures to Enhance Biodiversity: an Illustrated Guide.* Urbane Project, http://urbaneproject.org/assets/pdf/Firth-Image%20guide.pdf (accessed July 2018).

Fish, J. D. & Fish, S. 2011. *A Student's Guide to the Seashore*, 3rd edition. Cambridge University Press, Cambridge.

Flöthe, C. R. & Molis, M. 2013. Temporal dynamics of inducible anti-herbivory defences in the brown seaweed *Ascophyllum nodosum* (Phaeophyceae). *Journal of Phycology* 49: 468–474.

Gibbs, P. E., Pascoe, P. L. & Burt, G. R. 1988. Sex change in the female dog-whelk, *Nucella lapillus*, induced by tributyltin from antifouling paints. *Journal of the Marine Biological Association of the UK* 68: 715–731.

Gibson, R., Hextall, B. & Rogers, A. 2001. *Photographic Guide to the Sea and Shore Life of Britain and North-West Europe.* Oxford University Press, Oxford.

Gosse, P. H. 1854. *The Aquarium: an Unveiling of the Wonders of the Deep Sea.* John van Voorst, London.

Gosse, P. H. 1856. *Tenby: a Sea-Side Holiday.* John van Voorst, London.

Gosse, P. H. 1865. *A Year at the Shore.* Alexander Strahan, London.

Guiry, M. D. & Guiry, G. M. 2017. AlgaeBase. National University of Ireland, Galway. www.algaebase.org (accessed July 2018).

Hardy, A. 1956. *The Open Sea: its Natural History. Part I: The World of Plankton.* New Naturalist 34. Collins, London.

Hardy, A. 1959. *The Open Sea: its Natural History. Part II: Fish and Fisheries.* New Naturalist 37. Collins, London.

Harvey, W. H. 1846. *Phycologica Britannia, or a History of British Seaweeds*, Vols 1–3. Reeve Brothers, London.

Hayward, P. J. 1988. *Animals on Seaweed.* Richmond Publishing, Slough.

Hayward, P. J. 2004. *Seashore*. Collins, London.
Hayward, P. J. & Ryland, J. S. 1995. *Handbook of the Marine Fauna of North-West Europe*. Oxford University Press, Oxford.
Hoch, J. M., Schneck, D. T. & Neufeld, C. J. 2016. Ecology and evolution of phenotypic plasticity in the penis and cirri of barnacles. *Integrative and Comparative Biology* 56: 728–740. doi: 10.1093/icb/icw006.
Holdt, S. L. & Kraan, S. 2011. Bioactive compounds in seaweed: functional food applications and legislation. *Journal of Applied Phycology* 23: 543–597.
Honegger, R. 2000. Simon Schwendener (1829–1919) and the dual hypothesis of lichens. *The Bryologist* 103: 307–313.
Higgins, N. F., Connan, S. & Stengel, D. B. 2015. Factors influencing the distribution of coastal lichens *Hydropunctaria maura* and *Wahlenbergiella mucosa*. *Marine Ecology* 36: 1400–1414.
Jenkins, S. R., Hawkin, S. J. & Norton, T. A. 1999. Direct and indirect effects of a macroalgal canopy and limpet grazing in structuring a sheltered intertidal community. *Marine Ecology Progress Series* 188: 81–92.
Jones, W. E. 1962. A key to the genera of the British Seaweeds. *Field Studies* 1 (4): 1–32.
Juniper, T. 2013. *What Has Nature Ever Done for Us?* Profile Books, London.
Juniper, T. 2015. *What Nature Does for Britain*. Profile Books, London.
King, P. E. 1974. *British Sea Spiders*. Academic Press, London.
Kiørboe, T., Andersen, A., Langlois, V. & Jakobsen, H. H. 2010. Unsteady motion: escape jumps in planktonic copepods, their kinematics and energetics. *Journal of the Royal Society, Interface* 7 (52): 1591–1602. doi: 10.1098/rsif.2010.0176.
Kohlmeyer, J. & Volkmann-Kohlmeyer, B. 2003. Marine ascomycetes from algae and animal hosts. *Botanica Marina* 46: 285–306.
Kraberg, A., Bauman, M. & Durselen, C. 2010. *Coastal Phytoplankton: Photo Guide for Northern European Seas*. Verlag Dr. Friedrich Pfeil, Munich.
Lankester, E. R. 1910. *Science from an Easy Chair*. Methuen, London.
Larink, O. & Westheide, W. 2011. *Coastal Plankton: Photo Guide for European Seas*, 2nd edition. Verlag Dr. Friedrich Pfeil, Munich.
Lebour, M. V. 1930. *The Planktonic Diatoms of the Northern Seas*. Ray Society, London.
Leggett, M. C., Wilcockson, R. W., Day, V. D., Phillips, S. & Arthurs, W. 1996. The genetic effects of competition in seaweed flies. *Biological Journal of the Linnean Society* 57: 1–11.
Leliaert, F., Verbruggen, H., Vanormelingen, P. et al. 2014. DNA-based species delimitation in algae. *European Journal of Phycology* 49: 179–196.
Lewis, J. R. 1964. *The Ecology of Rocky Shores*. Hodder & Stoughton, London.
Lewis, J. R. & Bowman, R. S. 1975. Local habitat induced variations in the population dynamics of *Patella vulgata*. *Journal of Experimental Marine Biology and Ecology* 17: 165–203.
Little, C., Williams, G. A. & Trowbridge, C. D. 2009. *The Biology of Rocky Shores*, 2nd edition. Oxford University Press, Oxford.
Marshall, S. A. 2012. *Flies: the Natural History and Diversity of Diptera*. Firefly Books, Richmond Hill, Ontario.
Mason, T. H. 1938. *The Islands of Ireland*. Batsford, London.
McMeechan, F., Manica, A. & Foster, W. 2000. Rhythms of activity and foraging in the intertidal insect *Anurida maritima*: coping with the tide. *Journal of the Marine Biological Association of the United Kingdom* 80: 189–190.
Melzer, R. R. & Meyer, R. 2010. Field experiments on the association of decapod crustaceans with sea anemones, *Anemonia viridis* (Forsskål, 1775). *Natura Croatica* 19: 151–163.
Millar, R. H. 1970. *British Ascidians*. Academic Press, London.
Moore, P. G. & Seed, R. (eds). 1985. *The Ecology of Rocky Coasts: Essays Presented to J. R. Lewis*. Hodder & Stoughton, London.
Mouritsen, O. G. 2013. *Seaweeds Edible, Available and Sustainable*. University of Chicago Press, Chicago, IL.
Naylor, E. 1972. *British Marine Isopods: Keys and Notes for Identification of the Species*. Academic Press, London.
Naylor, P. 2005. *Great British Marine Animals*, 2nd edition. Sound Diving Publications, Plymouth.
Newell, R. 1979. *Biology of Intertidal Animals*. Marine Ecological Surveys, Faversham, Kent.
Newton, L. 1931. *A Handbook of the British Seaweeds*. British Museum (Natural History), London.
Nicolson, A. 2017. *The Seabird's Cry: the Lives*

and Loves of Puffins, Gannets and Other Ocean Voyagers. Collins, London.

Oldroyd, H. 1964. *The Natural History of Flies*. Weidenfeld & Nicolson, London.

Pearson, J. 1908. *Cancer* (the edible crab). *Liverpool Marine Biology Committee Memoirs* 16.

Pereira, D. M., Correia-da-Silva, G., Valentão, P., Teixeira, N. & Andrade, P. B. 2014. GC-MS lipidomic profiling of the echinoderm *Marthasterias glacialis* and screening for activity against human cancer and non-cancer cell lines. *Combinatorial Chemistry & High Throughput Screening* 17: 450–457.

Picton, B. E. 1993. *A Field Guide to the Shallow-water Echinoderms of the British Isles*. Immel, London.

Purvis, W. 2000. *Lichens*. Natural History Museum, London.

Rainbow, P. S. 1984. *An Introduction to the Biology of British Littoral Barnacles*. Field Studies Council, Shrewsbury. (Originally published in *Field Studies* 6: 1–51.)

Reid, D. G. 1993. Barnacle-dwelling ecotypes of three British *Littorina* species and the status of *Littorina neglecta*. *Journal of Molluscan Studies* 59: 51–62.

Russell, F. S. & Yonge, C. M. 1975. *The Seas*. Warne, London.

Sabourault, C., Ganot, P., Deleury, E., Allemand, D. & Furla, P. 2009. Comprehensive EST analysis of the symbiotic sea anemone, *Anemonia viridis*. *BMC Genomics* 10: 333.

Sampath-Wiley, P., Neefus, D. & Jahnke, L. S. 2008. Seasonal effects of sun exposure and emersion on intertidal seaweed physiology: fluctuations in antioxidant contents, photosynthetic pigments and photosynthetic efficiency in the red alga *Porphyra umbilicalis* Kützing (Rhodophyta, Bangiales). *Journal of Experimental Marine Biology and Ecology* 361: 83–91.

Sancho, L. G., de la Torre, R., Horneck, G. et al. 2007. Lichens survive in space: results from the 2005 LICHENS Experiment. *Astrobiology* 7: 443–454. doi:10.1089/ast.2006.0046.

Savoca, M. S., Wohlfeil, M. E., Ebeler, S. E. & Nevitt, G. A. 2016. Marine plastic debris emits a keystone infochemical for olfactory foraging seabirds. *Science Advances* 2 (11): e1600395. doi: 10.1126/sciadv.1600395.

Schmitt, W. L. 1973. *Crustaceans*. David & Charles, Newton Abbot.

Schonbeck, M. W. & Norton, T. A. 1979. An investigation of drought avoidance in intertidal fucoid algae. *Botanica Marina* 22: 133–144.

Sheehan, E. V. & Cousens, S. L. 2017. 'Starballing': a potential explanation for mass stranding. *Marine Biodiversity* 47: 617–618.

Southward, A. J. 2008. *Barnacles*. Field Studies Council, Shrewsbury.

Spribille, T., Tuovinen, V., Resl, P. et al. 2016. Bascidiomycete yeasts in the cortex of ascomycete macrolichens. *Science* 353: 488–492.

Stabili, L., Acquaviva, M. I., Biandolino, F. et al. 2014. Biotechnological potential of the seaweed *Cladophora rupestris* (Chlorophyta, Cladophorales) lipidic extract. *New Biotechnology* 31: 436–444.

Sterry, P. & Cleave, A. 2012. *Collins Complete Guide to British Coastal Wildlife*. Collins, London.

Thomas, D. 2002. *Seaweeds*. Natural History Museum, London.

Thompson, T. E. & Brown, G. H. 1976. *British Prosobranch Molluscs*. Academic Press, London.

Tittley, I. 2016. *A New Atlas of the Seaweeds of Kent: The Flora Past and Present with Summaries for Essex, London, Sussex and Pas de Calais*. Kent Field Club, Doddington.

Trewhella, S. & Hatcher, J. 2015. *The Essential Guide to Beachcombing and the Strandline*. Wild Nature Press, Plymouth.

Weinbauer, G., Nussbaumer, V. & Patzner, R. A. 1982. Studies on the relationship between *Inachus phalangium* Fabricius (Maiidae) and *Anemonia sulcata* Pennant in their natural environment. *Marine Ecology* 3: 143–150.

Wijgerde, T. 2016. Victorian pioneers of the marine aquarium. *Advanced Aquarist* 17 (2). www.advancedaquarist.com/2016/2/aafeature2 (accessed July 2018).

Wilson, C. M., Crothers, J. H. & Oldham, J. H. 1983. Realized niche: the effects of a small stream on sea-shore distribution patterns. *Journal of Biological Education* 17 (1): 51–58.

Wood, C. 2013. *Sea Anemones and Corals of Britain and Ireland*, 2nd edition. Wild Nature Press, Plymouth.

Abbreviations

AONB	Area of Outstanding Natural Beauty	NCMPA	Nature Conservation Marine Protected Area
BES	Ballantine's Exposure Scale	NNR	National Nature Reserve
CD	Chart Datum	OD	Ordnance Datum
CPR	Continuous Plankton Recorder	OSPAR Convention	Oslo/Paris Convention (for the Protection of the Marine Environment of the North-East Atlantic)
DEFRA	Department for Environment, Food & Rural Affairs		
DMS	Dimethyl Sulphide	Ramsar	Ramsar Convention on Wetlands of International Importance
DNA	Deoxyribonucleic Acid		
EEZ	Exclusive Economic Zone	RMS	Royal Mail Ship
EHWS	Extreme High Water of Spring Tides	RNA	Ribonucleic Acid
ELWS	Extreme Low Water of Spring Tides	RRS	Royal Research Ship
EMS	European Marine Site	RSPB	Royal Society for the Protection of Birds
FSC	Field Studies Council		
HMS	Her Majesty's Ship	SAC	Special Area of Conservation
JNCC	Joint Nature Conservation Committee	SAMS	Scottish Association for Marine Science
MarClim	Marine Biodiversity and Climate Change Project	SPA	Special Protection Area
		SSSI	Site of Special Scientific Interest
MBA	Marine Biological Association	TBT	Tributyltin
MCZ	Marine Conservation Zone	UKHO	United Kingdom Hydrographic Office
MHWN	Mean High Water of Neap Tides	UNESCO	United Nations Educational, Scientific and Cultural Organization
MHWS	Mean High Water of Spring Tides		
MLWN	Mean Low Water of Neap Tides	VMCA	Voluntary Marine Conservation Area
MLWS	Mean Low Water of Spring Tides	WFD	Water Framework Directive
MNR	Marine Nature Reserve	WSSD	World Summit on Sustainable Development
MPA	Marine Protected Area		
MSFD	Marine Strategy Framework Directive		

Species names

Common English and scientific names are listed in alphabetical order of the English names, where these exist. Some species, for example of algae, lichens and polychaete worms, only have scientific names. These are listed under the group name (for example, 'algae').

Algae –
   Bangiomorpha atropurpurea
   Bangiomorpha pubescens
   Choreocolax polysiphoniae
   Gracilaria spp.
   Lithophyllum spp.
   Macrocystis spp.
   Polysiphonia spp.
   Prasiola stipitata
   Sargassum spp.
   Trebouxia spp.
   Trentepohlia spp.
American Oyster *Crassostrea virginica*
Amphipods –
   Gammarus locusta
   Hyperia galba
Angelshark *Squatina squatina*
Arctic Cowrie *Trivia arctica*
Arctic Tern *Sterna paradisaea*
Arrow worms *Sagitta* spp.
Ascidians (sea-squirts) –
   Aplidium spp.
   Botrylloides leachii
   Botryllus schlosseri
Atlantic Cod *Gadus morhua*
Atlantic Herring *Clupea harengus*

Bacteria –
   Bacillus licheniformis
   Helicobacter pylori
Ballan Wrasse *Labrus bergylta*
Banded pincer weeds *Ceramium* spp.
Barnacles –
   Austrominius (Elminius) modestus
   Balanus crenatus
   Chthamalus montagui
   Chthamalus stellatus
   Perforatus perforatus
   Sacculina carcini
   Semibalanus balanoides
   Trypetesa lampas
   Verruca stroemia
Barrel Jellyfish *Rhizostoma pulmo*
Basking Shark *Cetorhinus maximus*
Beadlet Anemone *Actinia equina*
Beetles –
   Aleochara obscurella
   Cafius xantholoma
   Omalium spp.
   Staphylinus spp.
Black Lichen *Lichina pygmaea*
Black Squat Lobster *Galathea squamifera*
Black Tar Lichen *Verrucaria (Hydropunctaria) maura*
Black-footed Limpet *Patella intermedia*
Bladder Wrack *Fucus vesiculosus*
Bloody Henry *Henricia oculata*
Bloody Henry *Henricia sanguinolenta*
Blue-rayed Limpet *Patella pellucida*
Bootlace Weed *Chorda filum*
Bootlace Worm *Lineus longissimus*
Breadcrumb Sponge *Halichondria panicea*
Bristletail *Petrobius maritimus*
Brittle-stars –
   Ophiocomina nigra
Broad-clawed Porcelain Crab *Porcellana platycheles*
Brooding Cushion-star *Asterina phylactica*
Brown Rat *Rattus norvegicus*
Brown Tuning Fork Weed *Bifurcaria bifurcata*
Bryozoans –
   Bowerbankia imbricata
   Electra pillosa
   Membranipora membranacea
Buck's-horn Plantain *Plantago coronopus*
Bushy Rainbow Wrack *Cystoseira tamariscifolia*

Species names

Butterfish *Pholis gunnellus*
By-the-wind Sailor *Velella velella*

Candy Stripe Flatworm *Prostheceraeus vittatus*
Carrion Crow *Corvus corone*
Chameleon-prawn *Hippolyte varians*
Channel Wrack *Pelvetia canaliculata*
China Limpet *Patella ulyssiponensis*
Chink shells *Lacuna* spp.
Chipolata Weed *Scytosiphon lomentaria*
Chitons –
   Acanthochitona crinita
   Lepidochitona cinerea
Cockle *Cerastoderma edule*
Coconut *Cocos nucifera*
Coin Shell *Lasaea adansoni*
Common Brittle-star *Ophiothrix fragilis*
Common Cushion-star *Asterina gibbosa*
Common Cuttlefish *Sepia officinalis*
Common Green Branched Weed *Cladophora rupestris*
Common Grey Sea-slug *Aeolidia papillosa*
Common Hermit Crab *Pagurus bernhardus*
Common Jellyfish (Moon Jellyfish) *Aurelia aurita*
Common Limpet *Patella vulgata*
Common Mussel *Mytilus edulis*
Common Piddock *Pholas dactylus*
Common Prawn *Palaemon serratus*
Common Sea Urchin (Edible Sea Urchin) *Echinus esculentus*
Common Seal *Phoca vitulina*
Common Shore Crab *Carcinus maenas*
Common Starfish *Asterias rubens*
Common Sunstar *Crossaster papposus*
Common Whelk *Buccinum undatum*
Compass Jellyfish *Chrysaora hysoscella*
Copepods –
   Calanus finmarchicus
   Tigriopus brevicornis
Coral weeds –
   Corallina officinalis
Corkwing Wrasse *Symphodus melops*
Cormorant *Phalacrocorax carbo*
Crab's-eye Lichen *Ochrolechia parella*
Creeping Chain Weed *Catenella caespitosa*
Crevice Brittle-star *Ophiopholis aculeata*
Crinoids –
   Marsupites (extinct)
   Uintacrinus (extinct)
Crustaceans –
   Tanais spp.
Cuckoo Wrasse *Labrus mixtus*

Dabberlocks *Alaria esculenta*
Dahlia Anemone *Urticina felina*
Diatoms –
   Bacillaria paxillifer
   Coccolithus (*Coccosphaera*) *pelagica*
   Coscinodiscus spp.
   Odontella spp.
   Phaeocystis globosa
Dinoflagellates –
   Protoperidinium spp.
Dogwhelk *Nucella lapillus*
Dulse *Palmaria palmata*
Dunlin *Calidris alpina*

Edible Crab *Cancer pagurus*
Edible Periwinkle *Littorina littorea*
Egg Wrack *Ascophyllum nodosum*
Entoprocts –
   Barentsia spp.
European Cowrie (Spotted Cowrie) *Trivia monacha*

Fifteen-spined Stickleback *Spinachia spinachia*
Five-bearded Rockling *Ciliata mustela*
Flat periwinkle aggregate –
   Littorina fabalis (*mariae*)
   Littorina obtusata
Flies (Diptera) –
   Aphrosylus spp.
   Helcomyza ustulata
   Scathophaga litorea
Forest Kelp *Laminaria hyporborea*
Fruit Fly *Drosophila melanogaster*
Fulmar *Fulmarus glacialis*
Fungi –
   Alternoria spp.
   Aspergillus spp.
   Cladosporium spp.
   Didymella fucicola
   Didymella magnei
   Fusarium spp.
   Lulworthia fucicola
   Lulworthia kniepii
   Mucor spp.
   Mycophycias (*Mycosphaerella*) *ascophylli*
   Penicillium spp.
   Phycomelaina laminariae
   Saprolegnia spp.
   Stemphylium spp.

Gannet *Morus bassanus*
Gem Anemone (Wartlet Anemone) *Aulactinia verrucosa*

Rocky Shores

Ghost Shrimp *Caprella linearis*
Giant Kelp *Macrocystis pyrifera*
Glass Prawn *Palaemon elegans*
Goose Barnacle *Lepas anatifera*
Grape Pip Weed *Mastocarpus stellatus*
Great Auk *Pinguinus impennis* (extinct)
Great Black-backed Gull *Larus marinus*
Greater Pipefish *Syngnathus acus*
Green Leaf Worm *Eulalia clavigera*
Green Sea Fingers *Codium fragile* ssp. *fragile*
Green Tar Lichen *Verrucaria* (*Wahlenbergiella*) *mucosa*
Grey Heron *Ardea cinerea*
Grey Seal (Atlantic Seal) *Halichoerus grypus*
Grey Topshell *Steromphala cineraria*
Gribble *Limnoria lignorum*
Guillemot *Uria aalge*
Gutweed *Ulva* spp.

Hairy Crab *Pilumnus hirtellus*
Herring Gull *Larus argentatus*
Hildenbrand's red weed *Hildenbrandia* spp.
Honeycomb Worm *Sabellaria alveolata*
Hooded Crow *Corvus cornix*
Horned Wrack *Fucus ceranoides*
Hornwrack *Flustra foliacea*
Housefly *Musca domestica*
Hydroids –
 Dynamena pumila
 Hydractinia echinata
 Obelia geniculata

Irish Moss (Carragheen) *Chondrus crispus*
Isopods –
 Campecopea hirsuta
 Dynamene bidentata
 Gnathia maxillaris
 Idotea granulosa
 Jaera spp.
 Sphaeroma serratum

Jackdaw *Coloeus monedula*
Japanese Oyster *Crassostrea gigas*
Juicy Whorl Weed *Chylocladia verticillata*

Keelworms –
 Spirobranchus lamarcki
 Spirobranchus triqueter
Killer Whale *Orcinus orca*
Kittiwake *Rissa tridactyla*

Laver *Porphyra* spp.
Leathery Sea-squirt *Styela clava*
Lesser Black-backed Gull *Larus fuscus*
Lesser Octopus *Eledone cirrhosa*
Lichens –
 Anaptychia runcinata
 Caloplaca marina
 Caloplaca thallincola
 Caloplaca verruculifera
 Collemopsidium foveolatum (*Pyrenocollema halodytes*)
 Collemopsidium pelvetiae
 Lichina confinis
 Opegrapha cesareensis
 Opegrapha physciaria
 Opegrapha rupestris
 Verrucaria amphibia
 Verrucaria degelii
 Verrucaria ditmarsica
 Verrucaria halizoa (*microspora*)
 Verrucaria microspora
 Verrucaria mucosa
 Verrucaria prominula
 Verrucaria striatula
 Xanthoria aureola (*ectaneoides*)
 Xanthoria elegans
Light-bulb Ascidian *Clavelina lepadiformis*
Little Cuttlefish *Sepiola atlantica*
Lobster *Homarus gammarus*
Long-clawed Porcelain Crab *Pisidia longicornis*
Lumpsucker *Cyclopterus lumpus*

Manx Shearwater *Puffinus puffinus*
Map Lichen *Rhizocarpon geographicum*
Marbled Crab (Montagu's Crab) *Xantho hydrophilus*
Marine Centipede *Strigamia maritima*
Midges –
 Clunio marinus
Mites –
 Halolaelaps spp.
Montagu's Blenny *Coryphoblennius galerita*
Mossy Feather Weed *Bryopsis plumosa*

Noctule Bat *Nyctalus noctula*
Northern Sea Urchin *Strongylocentrotus droebachiensis*
Northern Starfish *Leptasterias muelleri*
Nudibranchs (sea-slugs) –
 Facelina auriculata
 Polycera elegans
 Polycera faeroensis

Species names

Oarweed *Laminaria digitata*
Onion Fly *Delia antiqua*
Otter *Lutra lutra*
Oystercatcher *Haematopus ostralegus*

Painted Topshell *Calliostoma zizyphinum*
Pea Crab *Pinnotheres pisum*
Peacock's Fan *Padina pavonica*
Pepper Dulse *Osmundea pinnatifida*
Pheasant Shell *Tricolia pullus*
Phoronids –
    Phoronis hippocrepia
Pied Wagtail *Motacilla alba*
Pill-woodlouse *Armadillidium vulgare*
Pink paint weeds Corallinaceae
Polar Bear *Ursus maritimus*
Polychaete worms (bristle-worms) –
    Neoamphitrite figulus
    Nephtys spp.
    Nereis fucata
    Phyllodoce spp.
    Tomopteris spp.
Portuguese Man O' War *Physalia physalis*
Protists –
    Eutintinnus spp.
Pseudoscorpions –
    Neobisium maritimum
Puffin *Fratercula arctica*
Purple Sandpiper *Calidris maritima*
Purple Sea Urchin *Paracentrotus lividus*
Purple Topshell *Steromphala umbilicalis*
Purse Sponge *Grantia compressa*

Rabbit *Oryctolagus cuniculus*
Raven *Corvus corax*
Razorbill *Alca torda*
Red Cushion-star *Porania pulvillus*
Red Deer *Cervus elaphus*
Red Fescue *Festuca rubra*
Red Fox *Vulpes vulpes*
Red Ripple Bryozoan *Watersipora subatra*
Rock Pipit *Anthus petrosus*
Roe Deer *Capreolus capreolus*
Rosy Feather-star *Antedon bifida*
Rough periwinkle aggregate –
    Littorina arcana
    Littorina compressa (*nigrolineata*)
    Littorina neglecta
    Littorina saxatilis

St Piran's Crab *Clibanarius erythropus*
Sand Mason *Lanice conchilega*

Sandhoppers –
    Orchestia gammarellus
    Talitrus saltator
Scarlet-and-Gold Cup-coral *Balanophyllia regia*
Sea anemones –
    Calliactis parasitica
Sea Aster *Aster tripolium*
Sea beans *Mucuna* spp.
Sea-gherkin *Pawsonia saxicola*
Sea-hare *Aplysia punctata*
Sea Ivory *Ramalina siliquosa*
Sea lettuce *Ulva* spp.
Sea-mouse *Aphrodita aculeata*
Sea Oak *Halidrys siliquosa*
Sea-orange *Suberites ficus*
Sea Plantain *Plantago maritima*
Sea Slater *Ligia oceanica*
Sea-spiders –
    Achelia spp.
    Anoplodactylus spp.
    Nymphon spp.
Sea urchins –
    Strongylocentrotus pallidus
Seaweed flies –
    Coelopa spp.
    Fucellia tergina
    Orygma luctuosum
    Thoracochaeta spp.
Serrated Wrack *Fucus serratus*
Seven-armed Starfish *Luidia ciliaris*
Shag *Phalacrocorax aristotelis*
Shanny (Blenny) *Lipophrys pholis*
Shore Clingfish *Lepadogaster purpurea*
Shore Sea Urchin *Psammechinus miliaris*
Sipunculids (peanut worms) –
    Golfingia macintoshii
    Golfingia vulgaris
    Sipunculus nudus
Slipper Limpet *Crepidula fornicata*
Small Brittle-star *Amphipholis squamata*
Small Periwinkle *Melarhaphe neritoides*
Snails –
    Rissoa parva
Snakelocks Anemone *Anemonia viridis*
Snakelocks Anemone Shrimp *Periclimenes sagittifer*
Sperm Whale *Physeter macrocephalus*
Spider crabs –
    Hyas araneus
    Inachus dorsettensis
    Maia squinado
Spiny Squat Lobster *Galathea strigosa*
Spiny Starfish *Marthasterias glacialis*

Rocky Shores

Spiral tubeworms –
   Spirorbis corallinae
   Spirorbis inornatus
   Spirorbis spirorbis
Spiraled Wrack *Fucus spiralis*
Sponges –
   Hymeniacidon perlevis
Sprat *Sprattus sprattus*
Springtail *Anurida maritima*
Starling *Sturnus vulgaris*
Strawberry Anemone *Actinia fragacea*
Sugar Kelp *Saccharina latissima*
Syllid worms –
   Autolytus spp.
   Eusyllis blomstrandi

Thong Weed *Himanthalia elongata*
Thornback Ray *Raja clavata*
Thrift *Armeria maritima*
Tompot Blenny *Parablennius gattorugine*
Toothed Topshell (Common Topshell) *Phorcus lineatus*

Tope *Galeorhinus galeus*
Topknot *Zeugopterus punctatus*
Tortoiseshell Limpet *Testudinalia testudinalis*
Tough Laver *Porphyra umbilicalis*
Trematodes –
   Parochis acanthus
Turban Topshell *Gibbula magus*
Turnstone *Arenaria interpres*

Velvet Swimming Crab *Necora puber*

Water fleas –
   Daphnia spp.
   Evadne nordmanni
Wireweed *Sargassum muticum*
Worm Pipefish *Nerophis lumbriciformis*
Wrack Fly *Coelopa frigida*
Wrack Siphon Weed *Vertebrata* (*Polysiphonia*) *lanosa*
Wrinkled Rock-borer *Hiatella arctica*

Yellow Scales Lichen *Xanthoria parietina*

Illustration credits

Bloomsbury Publishing would like to thank those listed below for providing illustrations and for permission to reproduce copyright material within this book. While every effort has been made to trace and acknowledge all copyright holders, we would like to apologise for any errors or omissions, and invite readers to inform us so that corrections can be made in any future editions.

1, 2–3, 6 © Julian Cremona; 8 © Gordon Waterhouse; 9, 11, 12, 14, 16, 17, 18 © Julian Cremona; 19 © Julian Baker (JB Illustrations); 20, 21 © Julian Cremona; 22 © Adam Burton/Alamy Stock Photo; 23, 24 © Julian Cremona; 25 © Graham Hunt/Alamy Stock Photo; 27 © Julian Cremona; 28, 29, 30 © John Archer-Thomson; 31 © Jeff Tucker/Alamy Stock Photo; 32 © Philip Gosse; 33 top © Julian Cremona, bottom reproduced with kind permission from Shell International Limited; 34 © Julian Cremona; 35 © Clare Cremona; 36, 37, 38, 39 © Julian Cremona; 40 © Kevin Schafer/Alamy Stock Photo; 41 © Julian Cremona; 43 © Holmes Garden Photos/Alamy Stock Photo; 44 © Julian Cremona; 45 © age fotostock/Alamy Stock Photo; 46 © T. H. Mason, reproduced with kind permission of Erica Murray; 48 © Arterra Picture Library/Alamy Stock Photo; 49 © Julian Cremona; 50 © John Archer-Thomson; 53 © Susan McIntyre; 54 © Julian Cremona; 55, 56 © John Archer-Thomson; 57 © Julian Cremona; 58 © John Archer-Thomson; 59 © Julian Baker (JB Illustrations), courtesy of Mark Burton; 60 © John Archer-Thomson; 62 © Julian Cremona; 64 © John Archer-Thomson; 65 © Clare Cremona; 66 © John Archer-Thomson; 67 © Julian Cremona; 69, 70 © John Archer-Thomson; 73 © Julian Cremona; 74 © Susan McIntyre; 75 © John Archer-Thomson; 77 © Julian Baker (JB Illustrations), adapted from Figure 1 in Ballantine (1961), reproduced with kind permission of the Field Studies Council; 78, 79 © John Archer-Thomson; 80 top © John Archer-Thomson, bottom © Julian Cremona; 81 © John Archer-Thomson; 82, 83, 84 © Julian Cremona; 85, 86, 87, 89, 90 © John Archer-Thomson; 91 © Shirley Hibberd; 92 ©John Archer-Thomson; 93 © Julian Cremona; 95, 96, 97, 98 © Julian Cremona; 99 © John Archer-Thomson; 100, 101, 102 © Julian Cremona; 103, 104, 106 © John Archer-Thomson; 107 © Mike Crutchley; 108 © John Archer-Thomson; 109 © Julian Cremona; 111, 112, 113 © John Archer-Thomson; 115, 116 © Julian Cremona; 117 © Chronicle/Alamy Stock Photo; 118, 120, 121, 123, 124, 125, 126, 127, 128, 129 © John Archer-Thomson; 130 © Julian Cremona; 132 © John Archer-Thomson; 134 © Julian Cremona, inset © John Archer-Thomson; 135 © William Harvey; 136 from the collections of the Natural History Museum, London; 137 © DEA/BIBLIOTECA AMBROSIANA/Getty Images; 139 © John Archer-Thomson; 143 © Julian Cremona; 145, 147 © John Archer-Thomson; 148 © Julian Cremona; 149, 150 © John Archer-Thomson, inset © Julian Cremona; 151 © John Archer-Thomson; 153, 154 © Anne Bunker; 156, 157 © John Archer-Thomson; 158 top, bottom right © John Archer-Thomson, bottom left © Julian Cremona; 160, 161 © Anne Bunker; 162 © John Archer-Thomson; 163 © Anne Bunker; 164, 168 © John Archer-Thomson; 170 © Julian Cremona; 172 © The Natural History Collections of the University of Edinburgh; 173 © Julian Cremona; 174 © Science & Society Picture Library/Getty Images; 175, 176 © Julian Cremona; 178 left © Julian Cremona, right © John Archer-Thomson; 180 © Julian Cremona; 181 © John Archer-Thomson; 182 © UniversalImagesGroup/Getty Images; 183 © John Archer-Thomson; 184, 185, 186

© Julian Cremona; 189 © Julian Cremona; 190, 193 © John Archer-Thomson; 194 top © Julian Cremona, bottom © Mike Crutchley; 195, 197, 200, 201 © John Archer-Thomson; 202 © Julian Cremona; 203 © John Archer-Thomson (Figure 10 in Archer-Thomson, 2016, reproduced with kind permission of the Field Studies Council); 204, 205, 206, 207, 208, 209 © John Archer-Thomson; 210 © Julian Cremona; 211, 213, 214, 215 © John Archer-Thomson; 216 top © Julian Cremona, bottom © Mark Papp; 217, 218 © John Archer-Thomson; 219 © Julian Cremona; 220 © John Archer-Thomson; 221 © Julian Cremona; 222, 224, 225, 226, 227, 228, 229, 230 © John Archer-Thomson; 232 top © John Archer-Thomson, bottom © Julian Cremona; 233, 235 © John Archer-Thomson; 236, 238, 239, 240 © Julian Cremona; 241 © Mike Crutchley; 242, 243, 244, 246, 247 © Julian Cremona; 248 © John Archer-Thomson; 250, 251, 252 © Julian Cremona; 253 top © Mike Crutchley, bottom © Julian Cremona; 254, 256, 257, 258, 259, 260, 261 © Julian Cremona; 262 © Nature Picture Library/Alamy Stock Photo; 263 © John Archer-Thomson; 264, 265, 266, 267 © Julian Cremona; 268 © Jacques Descloitres, MODIS Land Rapid Response Team, NASA/GSFC; 269, 270 © Julian Cremona; 272 left © Phil Newman, right © Mike Crutchley; 273, 275 © Mike Crutchley; 276, 277 © Julian Cremona; 278, 279 © Mike Crutchley; 280 © Julian Cremona; 282 left © Mike Crutchley, right © Julian Cremona; 283, 284 © Mike Crutchley; 286, 287, 288, 291, 293, 295 © Julian Cremona; 296 © John Archer-Thomson; 297 © Ian Ireland; 299, 300, 301, 302 © Julian Cremona; 303 © John Archer-Thomson; 304, 305 © Julian Cremona; 306 © Laurie Campbell Photography; 307, 308, 310, 311, 312, 313, 315, 317, 319, 321, 322, 323 © Julian Cremona; 324 © Philip Mugridge/Alamy Stock Photo; 325 © Julian Cremona; 326 © Ashley Cooper/Getty Images; 328 top © John Archer-Thomson (Figure 1 in Archer-Thomson, 2016, reproduced with kind permission of the Field Studies Council), bottom © Susan McIntyre; 329 top © Susan McIntyre (Figure 4A in Archer-Thomson, 2016, reproduced with kind permission of the Field Studies Council), bottom © Julian Cremona; 330 © John Archer-Thomson; 331 © Susan McIntyre (Figure 9 in Archer-Thomson, 2016, reproduced with kind permission of the Field Studies Council); 332 © Susan McIntyre (Figure 15 in Archer-Thomson, 2016, reproduced with kind permission of the Field Studies Council); 333 © John Archer-Thomson (Figure 10 in Archer-Thomson, 2016, reproduced with kind permission of the Field Studies Council); 335, 336 © John Archer-Thomson; 337 top © Julian Cremona, bottom © John Archer-Thomson; 340 © Susan McIntyre; 342, 343, 345, 348 © John Archer-Thomson.

Index

Page numbers in **bold** refer to illustrations.
Page numbers in *italics* refer to tables.

abiotic factors 57
  geology and
    topography 67–8
  light 62–3
  pH 68, 338–9
  temperature
    fluctuations 58–60,
    339–41
  variation in salinity
    60–2
  water loss 56–8
  wave action 63–6
Acanthochitona crinita 219
Achelia spp. 243
acidification 338–9
Acrothoracica 250
Actinia equina 87
  fragacea 341
Aeolidia papillosa 217–18
agars 167, 168–9
agricultural runoff 334–5
Alaria esculenta 47
Alca torda 290
Aleochara obscurella 321
algae 119–22, 131, 133
  algal exploitation
    133–5
alginates 39, **39**, 167
Alternoria spp. 314
Amphipholis squamata
  114, 230
amphipods 79, 252,
  255–6
Anaptychia runcinata 124,
  125
Anemone, Beadlet 87,
  106, 106–8, **107**,
  109, **109**, **217**, 218
  Dahlia 112, **112**, 177
  Gem 110–12, **111**
  Snakelocks 106, **108**,
    108–10, 276
  Strawberry 341
  Wartlet 110–12
Anemonia viridis 106
Angelshark 41

animal fodder 169
Anoplodactylus spp. **242**,
  243
Antedon bifida 234–5
Anthozoa 15
Anthropocene 347
Anthus petrosus 72
Anurida maritima 100–2,
  101, 241
Aphrodita 184, 185
  aculeata 20
Aphrosylus spp. 320
Aplidium spp. 182
Aplysia punctata 218
aquaria **91**, 91–2, 108
arachnids 241–3
Aran Islands 44–5, **45**
Archer-Thomson,
  Anthony 63–4
Archer-Thomson, John
  (JA-T) 7, **9**, 124, 230,
  234, 332, 334
  The Chronicles of Larry:
    Volumes 1 & 2 8
Arctica islandica 191
Ardea cinerea 295
Areas of Outstanding
  Natural Beauty
  (AONBs) 30, 48
Arenaria interpres 72
Aristotle 231
Armadillidium vulgare 253
Armeria maritima 52
arthropods 235, 237,
  267
  Cheliceriformes
    (arachnids and
    others) 241–3
  crustaceans 244–67
  insects and others
    240–1
  key characteristics
    238–40
Ascidian, Light-bulb 181,
  181, **235**
Ascomycetes 122

Ascophyllum nodosum 39
Aspergillus spp. 314
astaxanthin 102
Aster tripolium 52
Aster, Sea 52
Asterias rubens 89, 225
Asterina gibbosa 115, 229
  phylactica 92, 115, 229
Atkins, Anna *Photographs*
  of British Algae 137
Atlantic tsunami, 7000–
  6000 BC 40
Auk, Great 41
auks 290
Aulactinia verrucosa
  110–12
Aurelia aurita 270
Austrominius (*Elminius*)
  modestus **248**, 248–9
Autolytus spp. 187

Bacillaria paxillifer 275
Bacillus licheniformis 169
bacteria 131, 313
Balanophyllia regia 292
Balanus crenatus 249
Ballantine, Bill 75–6
Ballantine's Exposure
  Scale (BES) 76–7, **77**,
  84, 212
banded pincer weeds
  158–9
Bangiomorpha atropurpurea
  155
  pubescens 155
Barentsia spp. **189**
Barnacle, Goose 249,
  325
barnacles **70**, 70–1, 74,
  88, 245–50
  acorn barnacles 61–2,
    62, 71, 200, 246, 247,
    248, 249, 285, 333,
    341
  feeding **247**, 247–8
  larval form 279, 283

  stellate barnacles 71,
    341
Barrett, John 33, **33**
  Collins Pocket Guide to
    the Seashore 33
Bat, Noctule 324
BBC 342, 346
  BBC Wildlife 346
Bdellidae 129, 286
Beachy Head, Sussex 21,
  22, 49
Bear, Polar 324
Bedruthan Steps,
  Cornwall 29, **30**
Bembridge, Isle of Wight
  20, **20**, 21, 23, 49
Bifurcaria bifurcata 341
Binns, R. 260
biofilm 97, 98, **98**
biological zones 52–3, **53**
biotic factors 69
  competition 69–71
  predation 72
birds 289–92, 307
  birds feeding on rocky
    shores 294–5
  birds feeding on
    strandline 322–3
bivalves 21, 68, 128–9,
  191, 192, 220–1, 225,
  343
Black Shields 124, **125**
blennies 301
Blenny, Montagu's 299
  Tompot 299
Bloody Henry 227–8,
  228
Blue Peter 342
Blue Planet 346
Bose, Louis 279
Botrylloides leachii 182,
  217
Botryllus schlosseri 182,
  183, 217
Bowerbankia 179
  imbricata 241

Breidbach, Olaf 287
bristle-worms 183–4, **184**
    errant bristle-worms 184–2
    sedentary bristle-worms 188–9
Bristletail 85, 240, 241
British Museum (Natural History) *Seaweeds of the British Isles* 137
British Phycological Society *Seaweeds of the British Isles* 137
Brittle-star, Common 230–1
    Crevice 230
    Small 114, 230
brittle-stars 99, 114, 223, 224, 225, **230**, 230–1
brown seaweeds 86, 88, 89, 143–53
    life history 153–4
Bryopsis plumosa 165
Bryozoan, Red Ripple 336
bryozoans **80**, 179–81, 241
Buccinum undatum 343
Bunker, F. *Seaweeds of Britain and Ireland* 137
Burren, Galway Bay 43–4, **44**, 67, **67**, 270
Butterfish 112, **113**, 299–300
Buxton, John 100
By-the-wind Sailor 269, **269**

Cafius xantholoma 321
Calanus finmarchicus 286
Calidris alpina 295
    maritima 295
Calliactis parasitica 265
Calliostoma zizyphinum 89, 207
Caloplaca 53, 85, **85**, 86
    marina 126
    thallincola 126, **127**
    verruculifera 126
Campbell, A. *Hamlyn Guide to Seashores and Shallow Seas of Britain and Europe* 8
Campecopea hirsuta 128
Cancer pagurus 240, 258
Caprella linearis 255–6
caprellids **256**
Capreolus capreolus 324

Carcinas maenas 18, 249
Cardigan Bay 303, 343
Carpenter's Rule 275
carrageenans 167, 168
Carragheen 47, 158
Catenella caespitosa 58
cats 290
Centipede, Marine 241
Cephalopoda 302
Ceramium spp. 159
Cerastoderma edule 72
Ceredigion, Wales 94
Cervus elaphus 324
cetaceans 303
Cetorhinus maximus 45
Chameleon-prawn 104
Channel Islands 233
Chart Datum (CD) 53–4, 209, 248, 339
Cheliceriformes 241–3
chemical pollution 335
children **13**
chink shells 217
chitons 88, 99, 216, 219
Chondrus crispus 47
Chorda filum 17
Choreocolax polysiphoniae 148
Chrysaora hysoscella 270
Chthamalus 71, 86, 249
    montagui **70**, **248**
    stellatus **70**, **248**, 341
Ciliata mustela 299
Cladocera 286
Cladophora 165
    rupestris 165
Cladosporium spp. 314
Clavelina lepadiformis 181
Clavell, Sir John 26
Cleave, A. *Collins Complete Guide to British Coastal Wildlife* 8
Clibanarius erythropus 28
cliffs 289–92
    guano effect 292–4
climate change 338–41
Clingfish, Shore **300**, 301
Clunio marinus 100, **101**, 241, 320
Clupea harengus 28
cnidarians 177–8
Coccolithus (*Coccosphaera*) *pelagica* 273
Cockle 72
Coconut 312, **312**
Cocos nucifera 312
Cod, Atlantic 28, 301
Codium fragile ssp. *fragile* 342

Coelopa 316
    frigida 318
Collemopsidium 130
    foveolatum 130, **130**
    pelvetiae 130
Coloeus monedula 322
colonial filter feeders 179–83
competition 69–71
Connect 336
Connell, J. H. 70–1
conservation 344–6
Continuous Plankton Recorder (CPR) 274
Convention on Biodiversity 344
Cook, James 48
copepods 102, 250–1, **251**, 283, **286**
    holoplankton 286
coral weeds 96–7, 99, **99**, **133**, 135, 159
Corallina 96, 135
    officinalis 179
Corallinaceae 96
corals 177
Cormorant 41, 290, 291–2, 294, **297**
Cornwall 304
Corvus corax 322
    cornix 322
    corone 322
Coryphoblennius galerita 299
Coscinodiscus spp. **273**, 275
Cowrie, Arctic 217
    European (Spotted) 217
Crab, Broad-clawed Porcelain 263–4, **264**
    Common Hermit 265
    Common Shore 18–20, 62, **80**, 81, 221, 249–50, **250**, 259–60, **260**, **261**, 261–2, 343
    Edible 240, 258–9, **259**, 260–1, **267**
    Hairy 258
    Long-clawed Porcelain 263
    Marbled (Montagu's) 258, **258**
    Pea 221
    St Piran's 28
    Velvet Swimming **239**, 257–8

crabs 79, 80, 84, 89, 110, 202, 203
    hermit crabs 80, 250, 264–5, **266**
    larval form 282, **283**, **284**
Crane, Nicholas *The Making of the British Landscape* 41
Crangon spp. 267
Crassostrea gigas 342
    virginica 342
Cremona, Julian (JC) 7, **8**, 8–9, 256, 270
    Field Atlas of the Seashore 8
Crepidula fornicata 342
crinoids 234–5
Crobie, James 119
Crook, Anne 42
Crossaster papposus 228
Crothers, John 33, 82, 201, 205
Crow, Carrion 322
    Hooded 322
Crump, Robin 92, **92**, 116
crustaceans 79, 99, 244–5
    amphipods 255–6
    barnacles 245–50
    copepods 250–1
    decapods 257–67
    isopods 252–5
    Malocostraca 251–2
Cup-coral, Scarlet-and-Gold **228**, 292
Cushion-star, Brooding 92, 115–17, **116**
    Common 115, **116**, 116–17
    Red 229
cushion-stars 114–17, 224–5
cuttlefish 191, 301–2
Cuttlefish, Common 302, **302**
    Little 302
Cyclopterus lumpus 297
Cytoseira tamariscifolia 26

Dabberlocks 47, 84, **133**, 150–1, 341
Dale Fort Field Centre, Pembrokeshire 7–8, 9, 32–3, **33**, 35, 76, 127, 192, 208, 240, 289, 292
    research at Frenchman's Steps 327–34

Index

temperature fluctuations **59**, 59–60
Toothed Topshell population **340**
Dalyell, John Graham 107–8
Daphnia spp. 286
Darwin, Charles 31, 141, 174, 179, 245–6, 273, 279
  Living Cirripedia 246
Dawkins, Richard 141, 182
  The Ancestor's Tale 162
decapods 257–67
Deer, Red 324, **324**
  Roe 324
dehydration 56–8
Delage, Yves 249
Delia antiqua 319
desiccation 56–7, 73, 89
Devon 17, 83, 110
  Devon Wildlife Trust 28
Diatom, Sliding 275
diatoms 97, 99, 159, 165, 231, 241, 254, 273, 274–6
Didymella fucicola 314
  magnei 314
Dillehay, Tom 133–4
Dillenius, Johan Jacob 135
dimethyl sulphide (DMS) 337
dinoflagellates 276, 278
DNA 138–9, 141
Dobson, Frank *Lichens: an Illustrated Guide* 121, 127
Dogwhelk 23, 57, 66, 82, 88, 191, 200–5, 211, 212, 216, 220–1
  Dale Fort population 332–4
  radula **202**
dolphins 47, 303
Dorset 7, 25–7, 110, 208
  Dorset Wildlife Trust 7, 26
  Jurassic Coast 68
Drew-Baker, Kathleen Mary 167
Drosophila melanogaster 318
Dubois, Raphael 21
Dulse 45–6, **133**, 135, 158, 169, **313**, 314

Dulse, Pepper 88, **100**, 156, **157**
Dumpton Gap, Kent 23
Dunlin 295
Dynamena **189**
  bidentata 254
  pumila 178, **178**

echinoderms 84, 221, 223
  brittle-stars 230–1
  echinoderm structure 223–5
  feather-stars 234–5
  sea urchins 231–3
  sea-cucumbers 233–4
  starfish 225–9
Echinus esculentus 17, 231
Egglishaw, Henry 318
Electra pillosa **180**
Eledone cirrhosa 301
Elford, William 279
Engelmann, T. W. 142
England 42
  east coast 22–3
  Kimmeridge Bay, Dorset **25**, 25–7
  south coast 21–2
  west coast 27–30
epitoky 187, 284–5
Eshaness, Shetland 74
Essex 23, 342
EU 345
  EU Marine Strategy Framework Directive 344
eukaryotes 140–1
Eulalia clavigera 184
European Space Agency (ESA) 122
Eusyllis blomstrandi 187
Eutintinnus spp. **287**
Evadne nordmanni 286
Exclusive Economic Zones (EEZs) 344
Extreme High Water of Spring Tides (EHWS) 53
Extreme Low Water of Spring Tides (ELWS) 53

Facelina auriculata 199
Feather-star, Rosy 234–5, **235**
feather-stars 84, 223, 225, 234–5
Fescue, Red **51**, 52, 85
Festuca rubra 52

Field Studies Council (FSC) 32, 35, 116, 347
Fish, J.D. and Fish, S. *A Student's Guide to the Seashore* 8, 219, 225
fishes 84, 89, 112–14, 289, 296–302, 307
Flaherty, Robert 44–5
Flamborough Head, Yorkshire 23, **23**, 48, 49, 344
Flatworm, Candy Stripe 173, **173**
flatworms **173**, 173–4, 176, 177
Flustra foliacea 179
Fly, Fruit 318
  Kelp 318
  Onion Maggot 319
  Wrack 318–19, **319**, 320
fossils 22–3, 27, 122, 155, 225, 234, 237
Fox, Red 290
Fratercula arctica 40
freshwater on the shore 81–2
Friends of the Earth 345
Fucellia 316, 319–20
  tergina 320
fucoidans 169
Fucus **154**, 292
  ceranoides 82, **82**
  serratus 88, 135
  spiralis 84
  vesiculosus 53
  v. var. *linearis* 66, 88, 147
Fulmar 47, 48, 292, 337, **337**
Fulmarus glacialis 47
fungi 119–22, 131, 313–14
Fusarium spp. 314
future challenges 327
  climate change 338–41
  conservation 344–6
  habitat destruction and mitigation 338
  invasive species 341–3
  long-term prognosis 346–7
  pollutants 334–7

Gadus morhua 28
Galathea squamifera 26, 264
  strigosa 264

Galeorhinus galeus 41
Galway Bay 303
gammarids **257**
Gammarus locusta 255, 256
Gannet 34, 47, 289, 290, 292, **336**
  guga hunting 305
Garstang, Walter 48, 182
Gatty, Margaret 137, **137**
geology 67–8
Giant's Causeway, Antrim 42, 47
Gibbs, Peter 23
Gibbula magus 89
Gnathia maxillaris 253
Golfingia macintoshii 171
  vulgaris **172**
Gosse, Edmund 91
Gosse, Philip Henry 31–2, 91
  Aquarium, The 91
  Tenby: a Sea-Side Holiday 31, **32**, 279–80
  Year at the Shore, A 31
Gower Peninsula, Wales 68
Gracilaria spp. 134
Grant, Robert 174, **174**, 175
Grantia compressa 175
Grassholm, Pembrokeshire 34, 292
Great Silkie of Sule Skerry, The 305
green seaweeds 81, 86, 96, 162–5
  life history 163
Green, William Spotswood 42
Greenpeace 336
Grey Seals Protection Act 1914 305
Gribble 253
Griffiths, Amelia 137
Griffithsia 137
guano effect 292–4
Guillemot 25, 47, 289–90, 292
Guiry, Mike 137
Gull, Great Black-backed 290–1
  Herring 290, 291, **291**
  Lesser Black-backed 290
gulls 294
gutweed 74, 165, 166, 251–2, 335

363

habitat destruction and mitigation 338
Haeckel, Ernst 273, 274, 287
Haematopus ostralegus 72
Halacaridae 241
Halichoerus grypus 37, 303
Halichondria panicea 175
Halidrys siliquosa 342
Halolaelaps spp. 321
Hardy, Sir Alister 274
　The Open Sea 274
Hardy, Thomas 26
Harvey, William Henry 27, 135–6, 137
Hayward, Peter 31
　Animals on Seaweed 244
Hebrides 343
　Inner Hebrides **14**
　Outer Hebrides 39, 301, 309
Helcomyza ustulata 320
Helicobacter pylori 169
Henricia oculata 227
　sanguinolenta 228
Henson, Victor 273–4
Heron, Grey 295, **295**
Herring, Atlantic 28
Hiatella arctica 68
Higgins, Niall 129–30
Hildenbrand's red weed 293
Hildenbrandia spp. 293
Himanthalia elongata 43
Hippolyte varians 104, 301
Hoch, Matthew 248
holoplankton 286–7
Homarus gammarus 264
Hooker, Sir Joseph 269, 273
Hornwrack 179
Hoskins, W. G. *The Making of the English Landscape* 41
Housefly 319
Huxley, Thomas 28, 246
Hyas araneus 263
Hydractinia echinata 265, **266**
hydroids 84, 89, 99, 177, 178, 179
Hydropunctaria maura 128
Hymeniacidon perlevis 175, **176**
Hyperia galba 256

Idotea granulosa 254, **254**
Inachus 110
　dorsettensis 263, **263**
industrial pollution 335
insects 240–1
International Maritime Organization 333
invasive species 341–3
Ireland 42–7, 213
　seaweed harvesting 45–7
Islander Kelp 47
Islay **38**, 39
Isle of Wight 20, 208–9, 342
isopods 79, 128, 252–5

Jackdaw 322
Jaera spp. 252
jellyfish 177, 270–1
　box jellyfish 177
Jellyfish, Barrel 256
　Common (Moon) 270–1
　Compass 270, **270**
Joint Nature Conservation Committee (JNCC) 344
Jones, Eifion *Key to the Genera of British Seaweeds* 137
Journal of Marine Ecology 129
Juniper, Tony *What Has Nature Ever Done for Us?* 345
　What Nature Does for Britain 345
Jura, Scotland **38**, 38–9

keelworms 189, 281
kelp 74, 84, 89, 92, 133–4, **309**, 314
Kelp, Forest 152, 167
　Giant 143
　Sugar 47, **133**, **150**, 150–1, 314
Kemp, Klaus 275–6, **276**
Kerry 303
Kimmeridge Bay, Dorset **25**, 25–7, 93, 298
Kiørboe, Thomas 250
Kitching, Jack *The Biology of Rocky Shores* 42
Kittiwake 25, 47, 48, 289, 290, 292

Labrus bergylta 296
　mixtus 297

Lacuna spp. 217
Laminaria **153**, 243
　digitata 36
　hyperborea 152
Lamlash Bay, Isle of Arran 344
Lamouroux, J. V. F. 136
Landmark Trust 26
Lanice conchilega 17
Lankester, Ray 28–9, 171, **182**
　Science from an Easy Chair 182
Larus argentatus 290
　fuscus 290
　marinus 290
Lasaea adansoni 88, 129, 221
laver 134
　laver bread 166–7
Laver, Tough 155
Lecanora spp. 85, **85**
Lepadogaster purpurea 301
Lepas anatifera 249
Lepidochitona cinerea 219
Leptasterias muelleri 227
Lewis, Jack *The Ecology of Rocky Shores* 42–3, 44, 44
Lichen, Black 88, **128**, 128–9, **221**, 242
　Black Tar 73, **119**, 127, **127**, 128, 129–30, 213
　Crab's-eye 85, **85**, 124, **125**
　Green Tar 86, 87, 128, **128**, **129**, 129–30
　Map 122
　Yellow Scales **125**, 126, 130
lichens 52–3, 117, 119–22, 131
　black lichens 127–9
　greeny-black lichens 129–30
　grey, brown and white lichens **124**, 124–5
　lichen distributions 130
　light green lichens 123
　orange lichens 126
　other species 130
　yellow lichens 126
Lichina confinis 120
　pygmaea 88, 120, 128
light 62–3
Ligia oceanica 85, 253

Limnoria lignorum 253
Limpet, Black-footed **109**, 199, 339, 341
　Blue-rayed 199, 200, **200**, 216
　China 97, **194**, 199, 341
　Common **191**, 192–3, 339
　Slipper 342–3, **343**
　Tortoiseshell 24, 341
limpets 57, 66, 74, 86, 88, 96, **96**, 191, 192–200, 216
　fighting **69**
　Frenchman's Steps population 327–34, **328**, **329**, **331**, **332**
　radula 97, **97**, **194**
　reproduction 60, **60**, 195–6
　rock hardness 68
Lineus longissimus 172–3
Linnaeus, Carl 135, 245
Lipophrys pholis 113, 298
Lithophyllum spp. 314
Little Skellig, County Kerry **293**
Little, C. *The Biology of Rocky Shores* 8, 42, 71, 102, 159, 198
Littorina arcana 138, 211–12
　compressa (*nigrolineata*) **211**, 211, 212
　fabilis (*mariae*) 214, 215
　littorea 87, 213
　neglecta 212
　obtusata 214–15, 216
　saxatilis 138, 211, 212
Lobster 264
　Black Squat 26, 264
　Spiny Squat 264, **265**
locations of rocky shores around Britain and Ireland **18**
Lough Hyne, Cork 42, 83
Lovelock, James 141
Luidia ciliaris 228
Lulworthia fucicola 314
　kniepii 314
Lumpsucker 297–8
Lundy Island 344
Lutra lutra 37

Macrocystis 134
　pyrifera 143
maerl beds 38, **227**

Maia squinado 262, **262**
Malocostraca 251–2
mammals 289, 307
  mammals feeding on strandline 324
  marine mammals 303–6
Man of Aran 44–5
MarClim (Marine Biodiversity and Climate Change Project) 208, 340–1, 345
Margulis, Lynn 141
Marine and Coastal Access Act (UK) 2009 344
Marine Biological Association (MBA) 23, 28–9, 341
Marine Conservation Areas (MCAs) 47
Marine Conservation Society 336, 344, 347
Marine Nature Reserves 42
Marine Protected Areas (MPAs) 343, 344, 346
marine research 35
Marsupites spp. 23
Marthasterias glacialis 226–7
Mary Rose 342
Mason, Thomas *The Islands of Ireland* 45–6
mass extinction events 347
Mastocarpus stellatus 158
Maxillopoda 250, 251
Mean High Water of Neap Tides (MHWN) 53
Mean Low Water of Neap Tides (MLWN) 53
Melarhaphe neritoides 72, 212
Membranipora membranacea 179
Mereschkowski, Konstantin 140–1
mermaid's purses **310**
microhabitat availability 78–80
midges, marine 100, 241, 320

Milford Haven 13, 32, 34, 53–4, 73–4, 77, 83, 93, 278, 285, 335
Millport Marine Laboratory, Great Cumbrae, Scotland 35, **35**
mites 321
  marine mites 241, **241**, 242
Möller, Johann Diedrich 275
molluscs 191–2
  bivalves 220–1
  dogwhelks 200–5
  herbivorous guilds – who eats what? 216
  limpets 192–200
  periwinkles 210–15
  snails, slugs, cowries and chitons 217–19
  topshells 206–9
Monkstone Point, Pembrokeshire 225
Moore, P. G. *The Ecology of Rocky Coasts* 72, 144, 146, 195
Moray Firth 303
Morrell, Stephen 143
Morus bassanus 34
Moss, Irish 47, 96, **133**, 158
Motacilla alba 322
Mucor spp. 314
Mucuna spp. 312
Müller, Johannes 273
Müller, Otto 279, 281
Murray, Sir John 35
Musca domestica 319
Mussel, Common 61, 71, 82, 89, 200–1, 220, **220**, 343
mussels 57, 88, 191
  horse mussels 84
Mycophycias (*Mycosphaerella*) *ascophylli* 144, 314
mysids 251–2, 267
Mytilus edulis 61, 220

National Trust 28, 32, 49, 347
Nature Conservation Marine Protected Areas (NCMPAs) 344
nauplius larva 279, 283, **283**
Naylor, Paul *Great British Marine Animals* 234

neap tides 53
  neap tide strandlines 310, **310**
Necora puber 257
nematodes 172
Nemertea 172
Neoamphitrite 188
  figulus 17
Neobisium maritimum 88, 129, 242, **242**
Nereis fucata 265
Nerophis lumbriciformis 300
Newton, Lilly *A Handbook of the British Seaweeds* 137
Nicholls, Eliza 26
Nicholls, J. *Hamlyn Guide to Seashores and Shallow Seas of Britain and Europe* 8
Nicolson, A. *The Seabird's Cry* 294
non-exploitation zones 344–5
Northumberland 41, 48, 230
Nucella lapilus 23
nudibranchs 183, 217–18
nurdles 337, **337**
nutrient recycling 307, 309, 325
  bacteria and fungi 313–14
  beetles and mites 320–1
  dead organic matter 311–12
  sandhoppers 315–16
  seaweed flies 316–20
  top strandline predators 322–4
Nyctalus noctula 324
Nymphon spp. 243

Oak, Sea 342
Oarweed 36, 47, 84, 150–1, **151**, **168**
Obelia spp. 178, **178**, 279
Ocean Rescue campaign 346
Ochrolechia parella 124
octopi 191, 290–1
Octopus, Lesser (Curler) 301
Odontella spp. 275
oil pollution 28, 68, 117, 127, 285, 329–31, 334

Oldroyd, Harold 318
Oltmanns, Friedrich 142–3
Omalium spp. 321, **321**
Opegrapha 120
  cesareensis 124
  physciaria 120
  rupestris 120
Ophiocomina nigra 230
Ophiopholis aculeata 230
Ophiothrix fragilis 230
Orchestia gammarellus 315
Orcinus orca 303
ordnance 104, **104**
Ordnance Datum (OD) 54
Ordnance Survey Landranger (1:50,000) maps 16
Orkney 40–1, 169, 208, 228, 233, 342
Orwell, George 38–9
Oryctolagus cuniculus 290
Orygma 316
  luctuosum 317
osmoconformers 61–2
osmoregulators 62
Osmundea pinnatifida 88
Otters 37, 306, **306**
Oyster, American 342
  Japanese 342
Oystercatcher 72, 198, **289**, 294–5

paddle worms 185
Padina pavonica 24
Pagurus bernhardus 265
Palaemon **7**
  elegans **266**
  serratus 104, 266
Palmaria palmata 45
Pancratium maritimum 56
Parablennius gattorugine 299
Paracentrotus lividus 42
parietin 126
Parochis acanthus 202
Patella 24
  intermedia 199
  pellucida 199
  ulyssiponensis 97, 199
  vulgata 192, 199
Pawsonia saxicola 233
Peacock's Fan 24, 26
peanut worms 171
Pelvetia canaliculata 57
Pembrokeshire 7–8, 9, 24, 32–3, 34, 67, 68,

73, 75, **91**, 92, 126, 209, 213, 304, 324
Penicillium spp. 314
Perforatus perforatus 249, 341
Periclimenes sagittifer 110
Periwinkle, Edible 87, 88, 201, **206**, 213–14, **214**, 216, 339
  larval form 281, **282**
  radula **216**
Periwinkle, Small 72, 86, 212–13, **213**
periwinkles 26–7, 57, 74, 96, 191, 341
  flat periwinkles 87, 88, 214–15, **215**, 216
  rough periwinkles 85, 86, 88, 138, **210**, 210–12, **213**
petrels 290
Petrobius maritimus 85, 240
pH 68, 338–9
Phaeocystis globosa **277**, 278
Phalacrocorax aristotelis 41, 290
  carbo 41, 290
Phoca vitulina 303
Pholas dactylus 21
pholasin 21–2
Pholis gunnellus 113
Phorcus lineatus 87, 207
Phoronis hippocrepia **180**
phycobilins 155–6
Phycomelaina laminariae 314
phyla 14, 171, 172, 174, 177, 179, 182, 191, 223, 237, 271, 286
Phyllodoce spp. 185, **277**
Physalia physalis 177
Physeter macrocephalus 309
physical zones 53–5
phytoplankton **269**, 271, **272**, 276, 277–8
Piddock, Common **21**, 21–2, 68
Pill-woodlouse 253
Pilumnus hirtellus 258
Pinguinus impennis 41
pink paint weeds 96, 97, **99**, **133**, 159, 199
Pinnotheres pisum 221
Pipefish, Greater 112, 300
  Worm 300

Pipit, Rock 72, 295, **295**, 322
Pisidia longicornis 263
plankton 267, 269, 289
  diatoms and dinoflagellates 274–6
  history of plankton research 273–4
  jellies 270–1
  permanency, or holoplankton 286–7
  phytoplankton blooms **269**, 277–8
  planktonic larval forms 281
  supply-side ecology 285
  water sampling 271–2
  zooplankton and metamorphosis 279–85
Plantago coronopus 52
  maritima 52
Plantain, Buck's-horn 52
  Sea 52, 85
plastic pollution 336–7
Pliny the Elder 21, 134–5, 219
pollution 334–7
Polycera elegans 199
  faeroensis **235**
polychaetes 183–9, 284–5
Polysiphonia spp. 159, **160**
Porania pulvillus 229
Porcellana platycheles 263
Porphyra 134, **161**, 166, 167
  umbilicalis 155
porpoises 303
Portuguese Man O' War 177, 269
post-Ice Age Britain 40–1
Prasiola stipitata 293
Prawn, Common 104, 266
  European Glass **266**
prawns 105, 266–7, 283
predation 72
problems for rocky-shore inhabitants 56
proboscis worms 172
Prostheceraeus vittatus 173
protists 287
Protoperidinium spp. **278**
Psammechinus miliaris 89, 232

pseudoscorpions 88, 242
Puffin 40, 47, 290, 291, 300
Puffinus puffinus 290
Pyrenocollema halodytes 130

Rabbit 290, 291
radiolarians 287
Rainbow, Phil *Introduction to the Biology of British Littoral Barnacles* 246
raised beaches 41, **41**
Raja clavata 41
Ramalina 123
  siliquosa 85, 123
Ramsar sites 344
Rat, Brown 290
Rathlin Island 47
Rattus norvegicus 290
Raven 322
Ray, Thornback 41
Razorbill 290
red seaweeds 58, 79, 88, 89, 92, 155–9
  life history 160–1
Reid, David 212
Renouf, Louis 42
Rhizocarpon geographicum 122
Rhizostoma pulmo 256
ribbon worms 172
Rio Earth Summit 1992 344
Rissa tridactyla 25
Rissoa parva 217
Robertson, David 35
Robin Hood's Bay, Yorkshire 48–9, **49**
rock pools 91–2
  baited lines 104
  biofilm 98
  brittle-stars, sea urchins and cushion-stars 114–17
  diversity of life in a pool 103–4
  fishes 112–14
  microscopic world 99–102
  pool specialists 96–7
  rock-pool conditions 92–5
  sea anemones 106–12
  shrimps and prawns 105
Rock-borer, Wrinkled 68
Rockall 47

Rockling, Five-bearded 299
rotifers, marine 286
rove beetles 315, 316, 320–1
Royal Botanic Gardens, Edinburgh 108
Royal Society 273
Royal Society for the Protection of Birds (RSPB) 34, 47, 347
Russell, Bernard 135
Ryland, John 31

Sabellaria alveolata 94
Saccharina latissima 47
Sacculina carcini **80**, 249–50, **250**
Sagan, Dorion 141
Sagitta spp. 286
salinity 60–2, 176
Sampath-Wiley, Priya 155
Sancho, Leopoldo 122
sand eels 292, 300
Sand Mason 17, 187, 284–5
sandhoppers 255, **315**, 315–16
Sandpiper, Purple 295
Saprolegnia spp. **313**
Sargassum 134
  muticum 20
Savoca, Matthew 337
scale worms **185**, 185–7
Scathophaga litorea 320
Schwenender, Simon 119
Science 131
Scotland 17, 42
  east coast to the Humber 47–9
  highlands and islands 36–9
  post-Ice Age Britain 40–1
  seal and Gannet hunting 305
Scottish Association for Marine Science (SAMS) 35
Scottish Natural Heritage 305
Scytosiphon lomentaria 152
sea anemones 61, 88, 106–12, 176, 177
sea-angling **307**
sea beans 312
sea-cucumbers 223, 225, **233**, 233–4

Index

Sea Empress oil spill 68, 117, 127, 285, 329–31, 334
Sea Fingers, Green 342
Sea-gherkin 233
sea-gooseberries 271
sea-hare 89, 191, **218**, 218–19
Sea Ivory 85, **85**, **119**, 123, **123**, 130
sea lettuce 74, 165, 166, 251–2, 335
sea-lilies 223, 234
sea-mats 70, 84, **84**, 89, 99, 179
Sea-mouse 20
Sea Orange 175
Sea-slug, Common Grey **217**, 217–18
sea-slugs 89, 181, 191, 199
sea-spiders 99, 181, 183, **242**, 242–3
Sea-squirt, Leathery 336
sea-squirts 70, 80, 83, 84, **84**, 89, 182, 217
Sea Urchin, Common (Edible) 17, **224**, **225**, 231–2, **232**
  Northern 233
  Pale 233
  Purple 42, 114, **115**
  Shore 89, 232, **232**, 338–9
sea urchins 114, 223, 224, 225, 231–3
  larval form 281–2, **282**
seabird colonies 289–92
  guano effect 292–4
seal hunting 305
Seal, Common 303–4
  Grey (Atlantic) 37, 303, **303**, **304**, 304–5, **305**
seals 303–5
seaweed baths 167
seaweed bread twists 166
seaweed flies 315, 316–20, **317**
seaweeds 78–9, 133
  algal exploitation 133–5
  brown seaweeds 143–54
  compost composition 311–12
  green seaweeds 162–5
  red seaweeds 155–61

seaweed science 135–7
seaweeds for food and industry 166–9
what are seaweeds? 140–2
why are seaweeds different colours? 142–65
Seed, R. *The Ecology of Rocky Coasts* 72, 144, 146, 195
Semibalanus balanoides 61–2, **70**, 71, 200, 245, **248**, 249, 285, 333, 341
Sepia officinalis 302
Sepiola atlantica 302
sewage 335
Seward, William H. 294
Shag 41, 290, **291**, 292, 294
Shanny **109**, 113–14, 298–9, **299**
Shark, Basking 45
Shearwater, Manx 290
shearwaters 290, 292
Shell, Coin 88, 221, **221**, 242
  Pheasant 217
Shetland 36, 40–1, 204, 228, 233
Shrimp, Ghost 255–6
  Snakelocks Anemone 110
shrimps 105, 267, 283
  opossum shrimps **237**, 252
sipunculids 171, **172**
Sipunculus nudus 172
Sir Alister Hardy Foundation for Ocean Science, Plymouth 274
Skokholm, Pembrokeshire 120
Skomer 34, **34**, **51**, 124, 230, 304
  Skomer Marine Conservation Zone 289, **345**
  Wick 78, **78**, 292
Sky TV 346
Slater, Sea 85, 253–4
social media 346
Solent 18–20, 21, 24, 25, 30, 342
Special Areas of Conservation (SACs) 28, 34, 303, 336, 344

species distribution 24
species identification 138–9
Sphaeroma serratum 254–5
spider crabs 262–3
Spinachia spinachia 112
Spirobranchus lamarcki 189
  triqueter 281
Spirorbis 99
  corallinae 188
  inornatus 188
  spirorbis 188
Sponge, Breadcrumb 175
  Purse 175
sponges 57, **57**, 70, 80, 83, 84, **84**, 88, 174–7
Sprat 292
Sprattus sprattus 292
spring tides 53
spring tide strandlines 310, **310**
springtails, marine 100–2, **101**, 241
Squatina squatina 41
SSSIs (Sites of Special Scientific Interest) 344
Stackpole Head, Pembrokeshire 303
Stanford, E. C. C. 167
Staphylinus spp. 320–1
starfish 221, 223, 224, 225–9
Starfish, Common 89, 198, 223, **223**, 225–6, **226**, 343
  Northern 227
  Seven-armed 228, **229**
  Spiny 226–7, **227**
Starling 322, **323**
Stemphylium spp. 314
Sterna paradisaea 289
Steromphala cineraria 89, 207
  umbilicalis 87, 207
Sterry, P. *Collins Complete Guide to British Coastal Wildlife* 8
Stickleback, Fifteen-spined 112, **297**
stingers 177–8
strandlines 309–10, 325
  bacteria and fungi 313–14
  beetles and mites 320–1

compost composition – dead organic matter 311–12
neap tide and spring tide strandlines 310, **311**
refuse **327**
sandhoppers 315–16
seaweed flies 316–20
top strandline predators 322–4
Strigamia maritima 241
Strongylocentrotus droebachiensis 233
  pallidus 233
Sturnus vulgaris 323
Styela clava 336
Suberites ficus 175
Sunstar, Common 228, **229**
syllid worms **171**, **186**, 187, 284, **284**
Symphodus melops 235, 297
Syngnathus acus 112

Talitrus saltator 315
Tanais spp. **238**
temperature fluctuations **59**, 59–60, 339–41
Tephromela atra 124
Tern, Arctic 289
terns 290, 292
Testudinalian testudinalis 24
Thalassiosira spp. 274
thermal discharges 335–6
Thompson, John 245
Thoracochaeta spp. 316
Thrift 52, 85, 295
tidal ranges 30
  effect of tidal range 72–3
  tidal range for selected sites in the British Isles 55
tidal rapids **37**, 37–9, 83–4
tides 51–2
  neap tide and spring tide strandlines 310, **310**
tidal heights and physical zones 53–5
Tigriopus brevicornis 102, **102**
tintinnids 287
Tittley, I. *A New Atlas of the Seaweeds of Kent* 23

Tomopteris spp. **280**
Tope 41
Topknot 301, **301**
topography 67–8
Topshell, Common 207
　Flat 207
　Grey 89, 207, 208–9
　Painted 89, **207**
　Purple 87, 88, **96**, **103**, 207, 208–9, 341
Topshell, Toothed 87, 88, 201, **206**, 207, 208–9, 339–40, 341
　radula **216**
Topshell, Turban 89
topshells 57, 74, 96, 191, 206–9, 216
Torrey Canyon disaster 28
Trebouxia spp. 120, 126
Trentepohlia spp. 120, 124
tributylin (TBT) 205, 333, 334
Tricolia pullus 217
Trivia arctica 217
　monacha 217
Trowbridge, C. D. *The Biology of Rocky Shores* 8
Trypetesa lampas 250
tubeworms **84**, 99, 188–9, **189**
tunicates 99, 181
Turnstone 72, 295, 322, **322**
Tyrian purple 205–6

Uintacrinus spp. 23
Ullinish, west Scotland **10**
Ulva spp. 81, 82, 86, 96, 100, 103, **133**, 162, **163**, 165, 166, 218, 251–2, 335
UN Convention on the Law of the Sea 344
UNESCO World Heritage Sites 42
Uria aalge 25
Ursus maritimus 324
Urticina felina 112, **112**

van Leeuwenhoek, Antonie 273
Velella velella 269
Verrucaria 53, 73, 85, 86, 120
　amphibia 127
　degelii 127–8
　ditmarsica 127
　halizoa (*microspora*) 127
　maura 127, 128, 213
　microspora 127
　mucosa 86, 127, 128
　prominula 128
　striatula 127, 128
　stroemia 249
Vertebrata (*Polysiphonia*) *lanosa* 79
vertical movement 51–2
Voluntary Marine Conservation Areas (VMCAs) 28
Vulpes vulpes 290

Wagtail, Pied 322, **323**
Wahlenbergiella mucosa 128
Wales 30–4, 42
Wallich, G. C. 273
Watchet Harbour, Somerset 82
water loss 56–8
water sampling 271–2
Watersipora subatra 336
wave action **16**, **17**, 63–6
　aspect 65
　depth of water offshore 65–6
　effect of wave exposure 73–7, **74**
　fetch 64–5
　measuring 74–5
　position 65
websites 47, 341, 344, 346
Weed, Bootlace 17
　Brown Tuning Fork 341
　Chipolata 152
　Common Green Branched **101**, **164**, 165

Creeping Chain 58, **58**, 86, 128, 156, **156**
Grape Pip 158–9, 293
Juicy Whorl 158
Mossy Feather **164**, 165
Thong 43, 88–9, 150, **150**
Wrack Siphon 79–80, **80**, **148**, 148–9
Wembury, Devon 28, **28**
Whale, Killer 303
　Sperm 309–10, **310**
Whelk, Common 343, **343**
Wildlife Trusts 344, 347
Williams, G. A. *The Biology of Rocky Shores* 8
Wireweed 20, **20**, **27**, 143, 341, 342
Wong, Yan *The Ancestor's Tale* 162
Worm, Bootlace 172–3
　Green Leaf 184
　Honeycomb 94
Worm's Head, Gower Peninsula 31, **31**
worms 99, 171
　arrow worms 286
　bristle-worms 184–9
　keelworms 189, 281
　leaf worms 185
　strange worms that are not worms 172–4
　true worm diversity 183–4
Wrack, Bladder 53, 66, **66**, 71, 88, 146, **147**, 147–8, 149, 152, 169, 214–15, **215**, 244, 312, 314, 339
　Bushy Rainbow **24**, 26
　Channel **56**, 57, 58, **58**, 84, 86, 130, 144–6, **145**, 156, 314
　Egg 39, 53, 79–80, **80**, 88, 143, **143**, **148**, 148–9, 167, 169, 214–15, 244, **306**, 312, 314

Horned 82
　Serrated 88–9, 96, 135, **135**, 149, **149**, 167, 178, 179, 188, 214, 215, 263
　Spiraled 84, 86, 146–7, **147**, 156
Wrasse, Ballan 296–7
　Corkwing 235, **296**, 297
　Cuckoo 297

Xantho hydrophilus 258
Xanthoria 53, 85, **85**, **119**, 120
　aureola (*ectaneoides*) **121**, 126
　elegans 122
　parietina 120, 126

yeasts 131

Zeugopterus punctatus 301
zonation 51, **54**, 72, **73**
　biological zones 52–3
　effect of tidal range 72–3
　effect of wave exposure 73–7, **74**
　freshwater on the shore 81–2
　microhabitat availability 78–80
　tidal heights and physical zones 53–5
　tidal rapids 83–4
　vertical movement 51–2
　walking down the shore 85–9
zooplankton and metamorphosis 279–85
zooxanthellae 106